MEASURE THEORY AND
FUNCTIONAL ANALYSIS

MEASURE THEORY AND FUNCTIONAL ANALYSIS

Nik Weaver

Washington University in St. Louis, USA

World Scientific

NEW JERSEY • LONDON • SINGAPORE • BEIJING • SHANGHAI • HONG KONG • TAIPEI • CHENNAI

Published by

World Scientific Publishing Co. Pte. Ltd.

5 Toh Tuck Link, Singapore 596224

USA office: 27 Warren Street, Suite 401-402, Hackensack, NJ 07601

UK office: 57 Shelton Street, Covent Garden, London WC2H 9HE

British Library Cataloguing-in-Publication Data
A catalogue record for this book is available from the British Library.

MEASURE THEORY AND FUNCTIONAL ANALYSIS

ISBN   978-981-4508-56-8

Printed in Singapore

Preface

This book is based on a set of notes I developed over several years of teaching a graduate course on measure theory and functional analysis. Its focal point is the stunning interplay between topology, measure, and Hilbert space exhibited in the spectral theorem and its generalizations. The prerequisites are minimal: readers need to be familiar with little beyond metric spaces and abstract real and complex vector spaces.

I have striven to eliminate unnecessary generality. Thus, whenever possible I assume topological spaces are metrizable, measure spaces are σ-finite, Banach spaces are either separable or have separable preduals, and so on, if there is any advantage in doing so. My rationale is that the objects of central importance in the subject all seem to be, in various senses, essentially countable, whereas the essentially uncountable setting houses a raft of pathology of no obvious interest. There are other benefits, as well: the machinery of generalized convergence (i.e., nets and filters) becomes largely superfluous, and appeals to the axiom of choice can generally be weakened to countable choice or even dropped altogether. I wonder how many analysts realize that the Hahn-Banach theorem, famous for its nonconstructive nature, requires no choice principle at all in the setting of separable Banach spaces.

Expert readers will notice numerous minor innovations throughout the book. Perhaps the most fruitful original idea is my incorporation of Hilbert bundles into the spectral theorem, a device I introduced in my book *Mathematical Quantization* (CRC Press, 2001). When I was a graduate student a friend advised me that the multiplication operator version of the spectral theorem is the form you understand, but the spectral measure version is the form you use. This is a pithy way of pointing out that although multiplication operators are more intuitive than spectral measures, they appear in spectral theory in a noncanonical and therefore somewhat inelegant manner. The Hilbert bundle approach neatly resolves this dilemma. Using only the elementary notions of Hilbert space direct sums and tensor products, one is able to formulate a more canonical multiplication operator version of the spectral theorem which, moreover, transparently exhibits both the underlying spectral measure and its multiplicity. Even more benefits accrue when we generalize spectral theory to families of commuting operators: the standard structure theorems for concrete abelian C*- and von Neumann algebras are augmented with spatial information which not only tells us that such algebras are abstractly isomorphic to $C_0(X)$ and $L^\infty(X)$ spaces, but also cleanly exhibits the way these abstract spaces are situated within $\mathcal{B}(H)$.

I wish to express my gratitude to all of my students who took this course over the past several years. Those were some very talented classes, and teaching them was a real pleasure.

This work was partially supported by NSF grant DMS-1067726.

Nik Weaver

When I'm working on a problem, I never think about beauty, I think only how to solve the problem. But when I have finished, if the solution is not beautiful, I know it is wrong.

— *Buckminster Fuller*

Contents

Chapter 1

Topological Spaces

1.1 Countability

We adopt the convention that 0 is not a natural number; thus $\mathbf{N} = \{1, 2, 3, \ldots\}$.

Definition 1.1. A set is *countably infinite* if there is a bijection between it and $\mathbf{N}$. It is *countable* if it is either finite or countably infinite. It is *uncountable* if it is not countable.

Countability conditions of various types will be assumed liberally throughout this book. Assuming that a set is countable can be very convenient because this means that its elements can be indexed as (a_n), with n ranging either from 1 to N for some N or from 1 to ∞, in either case giving us the ability to deal with them sequentially. Actually, the hypotheses we impose usually will not assert that the main set of interest is itself countable, but rather that in some way its structure is determined by a countable amount of information. This informal comment might make more sense after we discuss separability and second countability in Section 1.4.

Clearly $\mathbf{N}$ is countably infinite, since it is trivially in bijection with itself. The set of even natural numbers is also countably infinite via the bijection $n \leftrightarrow 2n$, and as the set of odd natural numbers is obviously in bijection with the set of even natural numbers, it is countably infinite too.

This shows that a countably infinite set (the natural numbers) can be split up into two countably infinite subsets (the even numbers and the odd numbers). Conversely, with a moment's thought it also shows that the union of two disjoint countably infinite sets will again be countably infinite: we can put one set in bijection with the even numbers and the other in bijection with the odd numbers, and then combine the two maps to establish a bijection between the union of the two sets and $\mathbf{N}$. For instance, we can use this idea to show that the set of integers $\mathbf{Z}$ is countably infinite. Define $f : \mathbf{Z} \to \mathbf{N}$ by

$$f(n) = \begin{cases} 2n & \text{if } n > 0 \\ -2n + 1 & \text{if } n \leq 0; \end{cases}$$

this is a bijection that matches the positive integers with the even natural numbers and the negative integers and zero with the odd natural numbers.

Next we observe that subsets and images of countable sets are always countable.

Proposition 1.2. *Let A be a countable set.*

(a) Any subset of A is countable.

(b) Any surjective image of A is countable.

Proof.   (a) We take it as known that any subset of a finite set is finite, so assume A is countably infinite. Let $f : \mathbf{N} \to A$ be a bijection and let B be any subset of A. If B is finite we are done, so assume B is infinite. Then $f^{-1}(B)$ must be an infinite subset of $\mathbf{N}$, so it has a smallest element, a second smallest element, etc. Let n_1 be the smallest element of $f^{-1}(B)$, n_2 the next smallest, and so on; then the map $k \mapsto f(n_k)$ is a bijection between $\mathbf{N}$ and B. So B is countably infinite.

(b) Suppose $f : A \to B$ is a surjection. Create a map $g : B \to A$ by, for each $b \in B$, letting $g(b)$ be an arbitrary element of $f^{-1}(b)$. Then g is a bijection between B and a subset of A, and it follows from part (a) that B must be countable. $\qquad\square$

This proposition illustrates why it is helpful to have a special term ("countable") for sets which are either finite or countably infinite. It is not true that any subset of a countably infinite set is countably infinite, nor is it true that any surjective image is countably infinite.

Having said that, in analysis the unqualified word "sequence" usually means "infinite sequence", i.e., a sequence indexed by $\mathbf{N}$, and we will follow this convention. If we want to consider finite sequences we will explicitly use the qualifier "finite".

Earlier we used the fact that the natural numbers can be partitioned into even numbers and odd numbers in order to show that the union of two disjoint countably infinite sets is always countably infinite. This result can be strengthened to say that a union of countably many sets, each of which is countable, will always be countable. We can show this by partitioning the natural numbers into odd numbers, odd multiples of 2, odd multiples of 4, odd multiples of 8, etc.

Proposition 1.3. *Any countable union of countable sets is countable.*

Proof. Let (A_n) be a finite or infinite sequence of countable sets. We must show that $\bigcup A_n$ is countable. We can assume that the sets A_n are disjoint; otherwise replace A_2 with $A_2 \setminus A_1$, A_3 with $A_3 \setminus (A_1 \cup A_2)$, etc. This makes the sets disjoint without affecting their union, and they stay countable by Proposition 1.2 (a).

Now for each n find an injection $f_n : A_n \to O$, where O is the set of odd natural numbers. Then define $f : \bigcup A_n \to \mathbf{N}$ by $f(a) = 2^{n-1} f_n(a)$ for $a \in A_n$. Since $f_n(a)$ is odd, $f(a)$ is divisible by 2^{n-1} but no higher power of 2. Thus the value of n can be read off from the power of 2 that appears in $f(a)$. It follows that if $f(a) = f(b)$ then a and b must belong to the same A_n, and then cancelling the 2^{n-1} factor yields $f_n(a) = f_n(b)$, and hence $a = b$, since f_n is injective. This shows that f is injective. So $\bigcup A_n$ is in bijection with a subset of $\mathbf{N}$, and hence it is countable by Proposition 1.2 (a). $\square$

The set of fractions with a given denominator q is in bijection with the set of integers via the correspondence $\frac{p}{q} \leftrightarrow p$. Since the set of possible denominators is countable, it follows that the set $\mathbf{Q}$ of rational numbers is a countable union of countable sets and hence is countable.

We also have the following corollaries.

Corollary 1.4. *Let A be a countable set and let $k \in \mathbf{N}$. Then the set A^k of all k-tuples of elements of A is also countable.*

Proof. We prove this by induction on k. It is trivial for $k = 1$. Now assume that A^k is countable for a given value of k and observe that for each $a \in A$ there is a natural bijection between A^k and the set of $(k+1)$-tuples of elements of A whose $(k+1)$st entry is a. Taking the union over a and applying Proposition 1.3 then shows that A^{k+1} is countable. By induction, we conclude that A^k is countable for all k. $\square$

For $k = 0$ we consider A^k to consist of a single element, an empty 0-tuple, so Corollary 1.4 is true in that case also.

Corollary 1.5. *The collection of all finite subsets of a countable set is countable.*

Proof.   Let A be a countable set. It will suffice to show that for each k the collection of k-element subsets of A is countable. Taking the union over k and applying Proposition 1.3 then yields the desired result.

We have a natural surjection from A^k onto the collection of k-element subsets of A given by $(a_1, \ldots, a_k) \mapsto \{a_1, \ldots, a_k\}$. We know from Corollary 1.4 that A^k is countable, so according to Proposition 1.2 (b) the collection of k-element subsets of A must also be countable, as desired.   □

Finally, we identify an example of an uncountable set.

Theorem 1.6. *The unit interval* $[0,1]$ *is uncountable.*

Proof.   Let (t_n) be any sequence in $[0,1]$; we will show that there is an element of $[0,1]$ not in the sequence.

Recursively define a descending sequence of closed intervals as follows. Start by letting $I_1 = [0,1]$. Having determined I_n, let I_{n+1} be either the closed left third or the closed right third of I_n, whichever choice will ensure that $t_n \notin I_{n+1}$. (Thus I_2 must be either $[0, \frac{1}{3}]$ or $[\frac{2}{3}, 1]$; if $I_2 = [0, \frac{1}{3}]$, say, then I_3 should be either $[0, \frac{1}{9}]$ or $[\frac{2}{9}, \frac{1}{3}]$; and so on.) Then the intersection $\bigcap I_n$ is nonempty; for example, the sequence of left endpoints of the intervals I_n is nondecreasing and bounded above, and hence it converges to a point t which evidently belongs to every I_n. For each n, since t_n does not belong to I_{n+1} it cannot equal t. Thus t is an element of $[0,1]$ that is not in the sequence (t_n).   □

The reason for using left and right thirds rather than left and right halves is that there is a possibility, however remote, that t_n is the exact midpoint of I_n. In that case both halves would contain t_n and we could not exclude it from I_{n+1}. Since the left and right thirds of I_n do not overlap, the proof given above avoids this minor difficulty.

We can infer from Theorem 1.6 and Proposition 1.2 (a) that the real line is also uncountable. If **R** were countable then any subset of it, in particular $[0,1]$, would also have to be countable.

1.2   Topologies

A topology can be thought of as the minimal structure needed to support a meaningful notion of continuity. Intuitively, a function is continuous if it takes nearby points in the domain to nearby points in the range; thus a continuous function that maps x to y should also map all the points that are sufficiently close to x to points that are sufficiently close to y. The essential ingredient here is not the quantitative evaluation of distances between points, but rather the qualitative notion of being close. This notion is formally captured by specifying which sets are "open". The idea is that a subset of a topological space is open if whenever it contains some point x, it also contains all the points sufficiently close to x.

We axiomatize this notion of openness in the following way.

Definition 1.7. A *topology* on a set X is a family $\mathcal{T}$ of subsets of X which satisfies

(i) $\emptyset, X \in \mathcal{T}$;
(ii) if $\{U_\alpha\}$ is any family of sets in $\mathcal{T}$ then $\bigcup U_\alpha \in \mathcal{T}$;
(iii) if $\{U_1, \ldots, U_n\}$ is a finite family of sets in $\mathcal{T}$ then $U_1 \cap \cdots \cap U_n \in \mathcal{T}$.

In words: $\mathcal{T}$ contains $\emptyset$ and X, and it is stable under arbitrary unions and finite intersections. Once X is equipped with a topology we call it a *topological space*.

A subset of X is said to be *open* if it belongs to $\mathcal{T}$ and *closed* if its complement belongs to $\mathcal{T}$. The *closure* of a set $A \subseteq X$, denoted $\overline{A}$, is the intersection of all closed sets containing A. A is *dense* in X if $\overline{A} = X$.

If every set that is open for one topology is also open for another, then the former topology (the one with fewer open sets) is said to be *weaker* and the latter (the one with more open sets) is said to be *stronger*. It will also be convenient to have available a minor variation on the concept of openness: a *neighborhood* of a point x is a subset of X that contains some open set containing x.

Any set can be endowed with a topology by declaring all subsets to be open. This trivial construction is actually important enough to merit a special term.

Example 1.8. Let X be a set. The *discrete topology* on X is the topology consisting of all subsets of X.

We will be most interested in topological spaces that arise in a natural way from a metric.

Example 1.9. Let X be a metric space with metric d. For $x \in X$ and $r > 0$ the *open ball* of radius r about x is the set

$$\text{ball}_r(x) = \{y \in X : d(x,y) < r\}.$$

The *metric topology* on X is defined by declaring that a subset of X is open if it contains an open ball of some positive radius about each of its elements. That is, U is open if for every $x \in U$ there exists $r > 0$ such that $\text{ball}_r(x) \subseteq U$.

The informal explanation of openness we gave above is based on this class of examples, glossing "U contains a ball of some positive radius about each of its elements" as "U contains all the points sufficiently close to each of its elements". This can help to motivate the definition of a topology since the metrically open sets are always stable under arbitrary unions and finite intersections, but not in general under, say, complements or countably infinite intersections.

Not every topology arises from a metric (which suggests that our informal explanation of openness may be misleading in some exotic cases). For instance, if X contains at least two elements then the topology for which $\emptyset$ and X are the only open sets clearly cannot arise from a metric. A simple condition that excludes uninteresting examples like this is the following.

Definition 1.10. A topological space X is *Hausdorff* if for any distinct $x, y \in X$ there exist disjoint open sets $U, V \subset X$ such that $x \in U$ and $y \in V$.

If this condition holds then we say that points can be separated by open sets. For our purposes topological spaces can generally be assumed to be Hausdorff.

We will now describe subobject, quotient, and product constructions which generate new topological spaces from old ones. The subobject construction is quite simple:

Definition 1.11. Let X be a topological space and let $A \subseteq X$. The *relative topology* on A is the topology consisting of all sets of the form $U \cap A$ for U an open subset of X.

Quotient spaces are also easy to define, although their nature is less transparent. For instance, it is not always easy to tell when a quotient space will be Hausdorff. Recall that the *quotient* of a set X by an equivalence relation $\sim$ is the set $X/\sim$ of equivalence classes for $\sim$. We denote the equivalence class of $x \in X$ by $[x]$.

Definition 1.12. Let $\sim$ be an equivalence relation on a topological space X. Then the *quotient topology* on $X/\sim$ consists of the sets $U \subseteq X/\sim$ with the property that $\pi^{-1}(U)$ is open in X, where $\pi : X \to X/\sim$ is the *quotient map* $\pi : x \mapsto [x]$.

The definition of product spaces is slightly more involved. Before we give this definition it will be helpful to introduce the notion of a base.

Definition 1.13. A family $\mathcal{B}$ of open sets in a topological space is a *base* for the topology if every open set is a union of sets in $\mathcal{B}$.

For the sake of brevity, we may also refer to a base for the topology on X as a base for X. The elements of a base are *basic open sets*.

This is an important concept because we often define a topology not by directly characterizing which sets are open, but by specifying a base. In practice, topological questions about such spaces are often recast in terms of the preferred base. Metric topologies can be viewed in this way:

Example 1.14. Let X be a metric space. Then the sets $\text{ball}_r(x)$ (for $x \in X$ and $r > 0$) constitute a base for the metric topology.

Now we define product topologies. Recall that the *cartesian product* of a finite sequence of sets $X_1, \ldots, X_N$ is the set $\prod X_n$ consisting of all N-tuples $\mathbf{x} = (x_n)$ with $x_n \in X_n$ for all n. If (X_n) is an infinite sequence of sets then $\prod X_n$ consists of all sequences $\mathbf{x} = (x_n)$ with $x_n \in X_n$ for all n.

Definition 1.15. If (X_n) is a finite or infinite sequence of topological spaces then the *product topology* on $\prod X_n$ is the topology consisting of arbitrary unions of sets of the form $\prod U_n$ where each U_n is an open subset of X_n and $U_n = X_n$ for all but finitely many n.

Thus the sets $\prod U_n$ constitute a base for the product topology. If it is not obvious that the product topology is actually a topology, this should be made clear by the discussion following Proposition 1.16 below.

The restriction "$U_n = X_n$ for all but finitely many n" is vacuous if the sequence (X_n) is finite. In that case we can simply say that a subset of the cartesian product is open if it can be written as a union of products of open sets. We will explain in the next section why the slight complication arises in the case of infinite products.

Alternatively, if we let $\pi_k : \prod X_n \to X_k$ be the kth coordinate projection defined by $\pi_k(\mathbf{x}) = x_k$, then the product topology can be described by saying it is the topology generated by the sets

$$\pi_k^{-1}(U) = \left\{ \mathbf{x} \in \prod_n X_n : x_k \in U \right\},$$

the set of points in the product space whose kth entry lies in U, where k is an index and U is an open subset of X_k. This is a valuable observation, so it is worth taking a moment to explain the notion of generation.

Proposition 1.16. *Let X be a set and suppose S is a family of subsets of X whose union is all of X. Then the family B of all finite intersections of sets in S is a base for a topology, namely the topology T consisting of all arbitrary unions of sets in B.*

Proof.   Once we have established that T (the family of arbitrary unions of sets in B) is a topology, the fact that B is a base for it is trivial. It is easy to see that T contains $\emptyset$ and X and is stable under arbitrary unions. To complete the proof we must show that T is stable under finite intersections. It will suffice to show that the intersection of any two sets in T is again in T. But any two sets in T will be of the form $\bigcup A_\alpha$ and $\bigcup B_\beta$ where the A_α and B_β belong to B, and their intersection can be written as

$$\left(\bigcup_\alpha A_\alpha \right) \cap \left(\bigcup_\beta B_\beta \right) = \bigcup_{\alpha,\beta} (A_\alpha \cap B_\beta).$$

Since each A_α and each B_β is a finite intersection of sets in S, the same is true of $A_\alpha \cap B_\beta$ for any α and β. This shows that the intersection of any two sets in T also belongs to T, and that completes the proof.   □

We say that the topology constructed in Proposition 1.16 is the topology *generated* by S.

Now given that the product topology is generated by the sets $\pi_k^{-1}(U)$, the fact that B is formed by taking only finite intersections of sets in S is what leads to the condition "$U_n = X_n$ for all but finitely many n" in the definition of the product topology. The point is that a finite intersection of sets of the form $\pi_k^{-1}(U)$ will consist of those elements of the product whose entries fall in specified open sets on some finite selection of coordinates and are otherwise arbitrary. In other words, it will be a set of the form $\prod U_n$ where each U_n is an open subset of X_n and $U_n = X_n$ for all but finitely many n.

There is one more general construction of topologies that we will use repeatedly.

Definition 1.17. Let X be a set and let $\{f_\alpha : X \to X_\alpha\}$ be a family of functions from X into topological spaces X_α. The *weak topology* on X determined by this family is the topology generated by the sets $f_\alpha^{-1}(U)$, letting α range over all indices and letting U range over all open subsets of X_α.

After we define continuity in the next section it will be clear that the weak topology is the weakest topology on X which makes all of the functions f_α continuous. It is generated by just those sets which need to be open in order for this to happen.

The product topology is a special case of the weak topology construction. Namely, it is the weak topology determined by the family of projection maps $\pi_k : \prod_n X_n \to X_k$ from the cartesian product into the factor spaces.

Relative topologies can be viewed as weak topologies too, although this is less interesting because they are so simple to begin with. If A is a subset of a topological space X, consider the inclusion map $\iota : A \hookrightarrow X$. The relative topology on A is the weak topology determined by this single function. For that matter, discrete and metric topologies can also be interpreted as weak topologies (exercise).

1.3 Continuous functions

Our intuition about continuity says that if a continuous function maps x to y then all the points sufficiently close to x must land sufficiently close to y. Thus, the inverse image of any neighborhood of y should contain a neighborhood of x; this condition would fail only if there were points close to x whose images were not close to y.

Since any open set is a neighborhood of all of its elements, we can characterize functions which are continuous at every point of their domain in terms of the global behavior of open sets:

Definition 1.18. Let X and Y be topological spaces. A function $f : X \to Y$ is *continuous* if the inverse image of any open subset of Y is an open subset of X.

It is a *homeomorphism* if it is bijective and both f and f^{-1} are continuous.

The appearance of inverse images in the defintion of continuity is convenient because inverse images interact well with set theoretic operations.

We have the easily verified laws

$$f^{-1}(A^c) = f^{-1}(A)^c$$
$$f^{-1}\left(\bigcup A_\alpha\right) = \bigcup f^{-1}(A_\alpha)$$
$$f^{-1}\left(\bigcap A_\alpha\right) = \bigcap f^{-1}(A_\alpha)$$

(inverse images commute with complementation, union, and intersection). Only the corresponding law for unions holds in general for pushforwards of subsets. This is related to the fact that having a subset of a set is equivalent to having a function from that set into $\{0,1\}$, namely, the characteristic function of the subset. So we can turn a subset of the range of f into a subset of the domain by composing its characteristic function with f. This makes the transformation of sets from range to domain "natural" in a way that pushing sets forward from domain to range is not.

In the metric space setting the open sets definition of continuity is equivalent to a quantitative version which explicitly requires that close points in the domain be mapped to close points in the range.

Proposition 1.19. *Let X and Y be metric spaces. Then a function $f : X \to Y$ is continuous with respect to the metric topologies on X and Y iff for every $x \in X$ and every $\epsilon > 0$ there exists $\delta > 0$ such that*

$$d(x,y) < \delta \qquad \Rightarrow \qquad d(f(x), f(y)) < \epsilon.$$

We leave the proof of Proposition 1.19 as an exercise. It helps to note that the displayed condition is equivalent to the condition

$$f(\mathrm{ball}_\delta(x)) \subseteq \mathrm{ball}_\epsilon(f(x)).$$

Next we present two basic results about continuity. The first is an alternative characterization in terms of closed sets which follows from the fact that inverse images commute with complementation, and the second states that the composition of two continuous functions is continuous.

Proposition 1.20. *Let X and Y be topological spaces and let $f : X \to Y$ be a function. Then f is continuous iff $f^{-1}(C)$ is closed in X for all closed $C \subseteq Y$.*

Proof. We have

$$f^{-1}(C) \text{ is closed for all closed } C \subseteq Y$$

if and only if

$$f^{-1}(U^c) \text{ is closed for all open } U \subseteq Y$$

if and only if

$$f^{-1}(U)^c \text{ is closed for all open } U \subseteq Y$$

if and only if

$$f^{-1}(U) \text{ is open for all open } U \subseteq Y,$$

as desired. □

Proposition 1.21. *Let X, Y, and Z be topological spaces and let $f : X \to Y$ and $g : Y \to Z$ be continuous functions. Then $g \circ f : X \to Z$ is continuous.*

Proof. Let U be an open subset of Z. Then $g^{-1}(U)$ is an open subset of Y since g is continuous, and hence $f^{-1}(g^{-1}(U))$ is an open subset of X since f is continuous. But

$$
\begin{aligned}
f^{-1}(g^{-1}(U)) &= \{x \in X : f(x) \in g^{-1}(U)\} \\
&= \{x \in X : g(f(x)) \in U\} \\
&= (g \circ f)^{-1}(U).
\end{aligned}
$$

Thus the inverse image under $g \circ f$ of any open set in Z is open in X. So $g \circ f$ is continuous. □

We now want to consider how continuity relates to the constructions discussed in Section 1.2. Rather than prove separate results for relative topologies, product topologies, and weak topologies, it is more efficient to start with weak topologies and then invoke the observation made at the end of Section 1.2 that relative topologies and product topologies are both special cases of weak topologies.

The proof of our result relating continuity to weak topologies uses the following generally helpful observation. Suppose the topology on Y is generated by a family $\mathcal{S}$. Then in order for a function $f : X \to Y$ to be continuous, it suffices that $f^{-1}(U)$ be an open subset of X for every $U \in \mathcal{S}$. That is because, as we know from Proposition 1.16, every open subset of Y is a union of finite intersections of sets in $\mathcal{S}$. Thus the inverse image of any open subset of Y will be a union of finite intersections of inverse images of sets in $\mathcal{S}$. So, appealing to the stability under arbitrary unions and finite intersections that is enjoyed by every topology, we see that if the inverse image of any set in $\mathcal{S}$ is open then the inverse image of any open set will be open.

Proposition 1.22. *Let $f : X \to Y$ be a map between topological spaces and suppose the topology on Y is the weak topology determined by a family* *of functions $g_\alpha : Y \to Y_\alpha$. Then f is continuous iff $g_\alpha \circ f$ is continuous for all α.*

Proof. The forward direction follows from Proposition 1.21 together with the fact that each g_α is continuous, which is immediate from the definition of the weak topology. Conversely, suppose $g_\alpha \circ f$ is continuous for all α. By the comment preceding the proposition, to show that f is continuous it will suffice to verify that $f^{-1}(g_\alpha^{-1}(U))$ is open for every index α and every open subset U of Y_α. But this follows immediately from our assumption that each $g_\alpha \circ f$ is continuous. $\qquad\square$

Corollary 1.23. *Let X be a topological space and let A be a subset of X equipped with the relative topology. Then a function from another topological space into A is continuous iff it is continuous when regarded as a map into X.*

Recall that $\pi_k : \prod X_n \to X_k$ denotes the projection map $\pi_k(\mathbf{x}) = x_k$.

Corollary 1.24. *Let X be a topological space, let (X_n) be a finite or infinite sequence of topological spaces, and let $f : X \to \prod X_n$ be a function. Then f is continuous iff $\pi_k \circ f : X \to X_k$ is continuous for all k.*

This last corollary should help explain the unexpected clause "$U_n = X_n$ for all but finitely many n" in the definition of the product topology. We already saw that this has to do with the product topology being generated by the sets $\pi_k^{-1}(U)$. The reason why this is a desirable generating set is because it produces the weak topology determined by the family of coordinate projections, and the payoff is that we get the universal property of product spaces described in Corollary 1.24.

Quotient spaces also enjoy a universal property. Say that a function $f : X \to Y$ *respects* an equivalence relation $\sim$ on X if $x \sim y$ implies $f(x) = f(y)$. Also observe that the quotient map $\pi : X \to X/\sim$ is continuous relative to the quotient topology on $X/\sim$. This is an immediate consequence of the definition of the quotient topology.

Proposition 1.25. *Let X and Y be topological spaces, let $\sim$ be an equivalence relation on X, and let $f : X \to Y$ be a continuous function that respects $\sim$. Then there is a continuous function $\tilde{f} : X/\sim \to Y$ such that $f = \tilde{f} \circ \pi$.*

Proof. Define $\tilde{f} : X/\sim \to Y$ by setting $\tilde{f}([x]) = f(x)$. This is well-defined because f respects $\sim$, and it is clear that $f = \tilde{f} \circ \pi$. To verify continuity, let $U \subseteq Y$ be open. Then $f^{-1}(U)$ is open in X since f is continuous, and since

$$\pi^{-1}(\tilde{f}^{-1}(U)) = f^{-1}(U)$$

it follows from the definition of the quotient topology that $\tilde{f}^{-1}(U)$ is open in $X/{\sim}$. This shows that $\tilde{f}$ is continuous. $\qquad\square$

1.4 Metrizability and separability

Although the axioms for topological spaces are simple and elegant, they are overly broad for our purposes. Consequently, we need to identify additional conditions which exclude behavior that is, as far as we are concerned, pathological and irrelevant. The two conditions that best suit our needs are metrizability and separability.

We already mentioned that we are mainly interested in topologies that arise from a metric. This motivates the next definition.

Definition 1.26. A topological space is *metrizable* if its topology is the metric topology for some metric.

At this point the reader might be wondering why we bothered to introduce the notion of a topological space in the first place, if we are only going to work with metric spaces. One answer is that nonmetrizable spaces will still play a role in our exposition. But a better answer is that even spaces which are in principle metrizable often do not come with a canonical metric. In such cases we are usually better off working at the level of topology and keeping in mind that the space could be equipped with a metric if necessary.

In Section 1.2 we discussed several methods for constructing topological spaces. Our first goal now is to show, to the extent we can, that these constructions produce metrizable spaces when given metrizable spaces as input. Our main result in this direction applies to weak topologies; since relative topologies and product topologies are special cases of weak topologies, it applies to them too.

Say that two metrics defined on the same set X are *equivalent* if they give rise to the same topology, and define the *diameter* of a metric d on X to be the (possibly infinite) quantity $\sup\{d(x,y) : x,y \in X\}$. We may also refer to this quantity as the diameter of X.

Lemma 1.27. *Let X be a metric space with metric d. Then there is an equivalent metric d' on X whose diameter is at most 1.*

Proof. Define

$$d'(x,y) = \min(d(x,y), 1)$$

for $x, y \in X$. It is easy to see that d' is also a metric. To verify equivalence it will suffice to check that every open ball for one metric is also an open set for the other metric. The balls of radius at most 1 for the two metrics are identical, so there is nothing to prove here. But that is enough because in any metric space any ball of radius greater than 1 is a union of balls of radius at most 1. □

Say that a family of functions $\{f_\alpha : X \to X_\alpha\}$ is *separating* if for every distinct $x, y \in X$ we have $f_\alpha(x) \neq f_\alpha(y)$ for some α.

Theorem 1.28. *The weak topology determined by any countable separating family of maps into metrizable spaces is metrizable.*

Proof. Let (f_n) be a finite or infinite separating sequence of functions from a set X into metrizable spaces X_n. For notational convenience, if this sequence is finite, fill it out to an infinite sequence by adding a string of functions into a space with one element. This obviously does not change the weak topology on X.

Fix a metric d_n on each X_n; by the lemma we can assume that each of these metrics has a diameter of at most 1. Then define a metric d on X by setting

$$d(x,y) = \sum_{n=1}^{\infty} \frac{d_n(f_n(x), f_n(y))}{2^n}.$$

The verification that this is a metric is straightforward. We need the f_n to be separating in order to ensure that $x \neq y$ implies $d(x,y) > 0$.

We will prove that the metric topology associated to d equals the weak topology determined by the f_n. First we show that any weakly open set is also metrically open. It is enough to check this for sets of the form $f_n^{-1}(U)$ where n is an index and $U \subseteq X_n$ is open, since these sets generate the weak topology. Thus fix n and U and let $x \in f_n^{-1}(U)$. Then $f_n(x) \in U$, and since U is open we must have $\text{ball}_\epsilon(f_n(x)) \subseteq U$ for some $\epsilon > 0$. It follows that if $d(x,y) < \frac{\epsilon}{2^n}$ then

$$d_n(f_n(x), f_n(y)) \leq 2^n \cdot d(x,y) < \epsilon,$$

and hence $f_n(y) \in \text{ball}_\epsilon(f_n(x)) \subseteq U$. This shows that $\text{ball}_{\epsilon/2^n}(x) \subseteq f_n^{-1}(U)$. As x was an arbitrary element of $f_n^{-1}(U)$, we conclude that this set is metrically open, as desired.

Conversely, to show that every metrically open set is weakly open, it will suffice to show that any ball about a point $x \in X$ contains a weakly open set containing x. Thus fix $x \in X$ and $\epsilon > 0$. Find N sufficiently large that $2^{-N} < \frac{\epsilon}{2}$; then we claim that the weakly open set

$$V = f_1^{-1}(\text{ball}_{\epsilon/2}(f_1(x))) \cap \cdots \cap f_N^{-1}(\text{ball}_{\epsilon/2}(f_N(x)))$$

contains x and is contained in $\text{ball}_\epsilon(x)$.

The first claim is easy. To see the second, let $y \in V$, so that $d_n(f_n(x), f_n(y)) < \frac{\epsilon}{2}$ for $1 \leq n \leq N$. Then

$$d(x, y) = \sum_{n=1}^{\infty} \frac{d_n(f_n(x), f_n(y))}{2^n}$$
$$< \sum_{n=1}^{N} \frac{\epsilon/2}{2^n} + \sum_{n=N+1}^{\infty} \frac{1}{2^n}$$
$$< \frac{\epsilon}{2} + \frac{\epsilon}{2} = \epsilon$$

and it follows that $y \in \text{ball}_\epsilon(x)$. We conclude that $x \in V \subseteq \text{ball}_\epsilon(x)$, as desired. $\qquad\Box$

The proof of Theorem 1.28 can give us some insight into weak topologies, and product topologies in particular. The natural base for the weak topology consists of finite intersections of sets of the form $f_\alpha^{-1}(U)$. Loosely speaking, only finitely many coordinates matter. The metric d used above replicates this effect by scaling down the target metrics as n increases, so that later coordinates matter less and less.

Corollary 1.29. *The relative topology on any subset of a metrizable space is metrizable.*

Corollary 1.30. *The product of a countable family of metrizable spaces is metrizable.*

Metrizability of quotient spaces is trickier. We will come back to this question later; see Corollary 3.56.

There is a second condition on topological spaces that we will often want to assume in addition to metrizability. It comes in two forms.

Definition 1.31. A topological space is *separable* if it contains a countable dense subset. It is *second countable* if the topology is generated by a countable family of open sets.

Note that if $\mathcal{S}$ is a countable family of open sets that generates the topology then the family $\mathcal{B}$ of finite intersections of sets in $\mathcal{S}$ is a base for the topology, and $\mathcal{B}$ is countable by Corollary 1.5. So every second countable space has a countable base.

In general second countability is stronger than separability, but for metrizable spaces the two are equivalent.

Proposition 1.32. *Let X be a metrizable space. Then X is separable iff it is second countable.*

Proof.   Suppose X is separable and let (x_k) be a dense sequence. Fix a metric on X and consider the family of sets $\text{ball}_{1/n}(x_k)$ with $k, n \in \mathbf{N}$. This is a countable family; we claim that it is a base for the topology, which will show that X is second countable.

To verify the claim, let $U \subseteq X$ be open and let $x \in U$; we must find n and k such that $x \in \text{ball}_{1/n}(x_k) \subseteq U$. First, find $m \in \mathbf{N}$ such that $\text{ball}_{1/m}(x) \subseteq U$. Then find $k \in \mathbf{N}$ such that $d(x, x_k) < \frac{1}{2m}$. Taking $n = 2m$ achieves the desired conclusion.

Conversely, suppose X is second countable and let (U_n) be a countable base for the topology. We can assume that each U_n is nonempty, so for each n choose an element $x_n \in U_n$. Since every open set is a union of sets in the base, it follows that every nonempty open set contains some x_n. This implies that (x_n) is dense in X, so we conclude that X is separable.   □

Separability may be more intuitive than second countability, but second countability is really the more natural condition. It relates more directly to the topology, which is defined in terms of subsets, not points.

In any case, within the setting of metrizable spaces we get the benefits of both points of view. This can be seen in the corollary to the next pair of results.

Proposition 1.33. *Any continuous image of a separable space is separable.*

Proof.   Let (x_n) be a dense sequence in a separable space X and let $f : X \to Y$ be a continuous surjection. Then $(f(x_n))$ is a sequence in Y, and if $U \subseteq Y$ were a nonempty open set that did not contain any $f(x_n)$ then $f^{-1}(U)$ would be a nonempty open subset of X that did not contain any x_n, a contradiction. Thus every nonempty open subset of Y contains some $f(x_n)$, so Y is separable.   □

Proposition 1.34. *The weak topology on a set X determined by any countable family of maps from X into second countable spaces is second countable.*

Proof. Let (f_n) be a finite or infinite sequence of functions from X into second countable spaces X_n. For each n let (U_k^n) be a sequence of open sets that generates the topology on X_n. Then the sets $f_n^{-1}(U_k^n)$ generate the weak topology on X. Thus the weak topology on X is countably generated, i.e., X is second countable. □

Corollary 1.35. *Let X be a separable metrizable space and let (X_n) be a finite or infinite sequence of separable metrizable spaces.*

(a) Any continuous image of X is separable.

(b) Any subset of X is separable.

(c) The product $\prod X_n$ is separable.

1.5 Compactness

In a discrete space all sets are open. At the opposite extreme are spaces in which the empty set and the whole space are the only open sets, but we are not interested in these spaces because they are not Hausdorff (provided the set contains more than one point). Within the Hausdorff setting, the property that naturally opposes discreteness is compactness. Discrete spaces have as many open sets as possible; in a sense that we will explain, subject to the Hausdorff condition compact spaces have as few open sets as possible.

The compactness property has several equivalent formulations. We start with the most standard one.

Definition 1.36. Let X be a topological space. An *open cover* of X is a family $\mathcal{O}$ of open subsets of X whose union equals X. A *subcover* of $\mathcal{O}$ is a subfamily $\mathcal{O}' \subseteq \mathcal{O}$ which is also an open cover. X is *compact* if every open cover of X has a subcover which contains only finitely many sets.

That is, whenever a family of open sets covers X, finitely many of them must already cover X. More briefly: every open cover has a finite subcover.

When we say that a subset C of X is compact we always mean with respect to the relative topology. This is equivalent to saying that every family of open subsets of X that covers C has a finite subfamily that also covers C. In other words, if we generalize the terminology of Definition 1.36 in the obvious way, so that an open cover of $C \subseteq X$ is a family of open subsets of X whose union contains C, etc., then the resulting definition of compactness for C is equivalent to its being compact for the relative topology.

The compactness condition is so strong that one might at first be surprised that it ever holds, outside of the trivial case where X is finite. Any finite space is compact because, given any open cover, for each point in the space we can select one set in the cover that contains it, and in this way extract a subcover that has at most as many sets as there are points in the space — fewer than this if there are any duplicate selections.

For an easy example of an infinite compact space, consider the set $\{0\} \cup \left\{ \frac{1}{n} : n \in \mathbf{N} \right\}$ with the relative topology it inherits from $\mathbf{R}$. Let $\mathcal{O}$ be an open cover. Then some $U \in \mathcal{O}$ contains 0, and since U is open it must also contain a ball about 0. But that means it contains the points $\frac{1}{n}$ for all sufficiently large n, or in other words, it excludes only finitely many points. Just as in the preceding paragraph, we can then find finitely many more sets in $\mathcal{O}$ which together with U cover the entire set. Thus, we have extracted a finite subcover from $\mathcal{O}$. So this space is compact.

A roughly similar idea appears in the following proof. Give $[0,1]$ its usual metric topology.

Theorem 1.37. $[0,1]$ *is compact.*

Proof. Let $\mathcal{O}$ be an open cover of $[0,1]$ and consider the set of points $t \in [0,1]$ with the property that the interval $[0,t]$ can be covered by finitely many sets in $\mathcal{O}$. Call this set A. Evidently if $t \in A$ and $s < t$ then $s \in A$. Let $t^* = \sup A$ and find $U_1 \in \mathcal{O}$ which contains t^*.

Clearly $t^* > 0$ since there is a set in $\mathcal{O}$ which contains 0 and therefore contains an interval to the right of 0. So U_1 contains an interval of the form $(t^* - \epsilon, t]$ for some $\epsilon > 0$. But $t^* - \epsilon \in A$, so there is a cover of $[0, t^* - \epsilon]$ by finitely many sets $U_2, \ldots, U_n \in \mathcal{O}$. If $t^* < 1$ then $\bigcup_{k=1}^{n} U_k$ contains the interval $[0, t^* + \delta]$ for some $\delta > 0$, contradicting the definition of t^*. So $t^* = 1$ and $U_1, \ldots, U_n$ cover $[0,1]$. Thus every open cover of $[0,1]$ has a finite subcover. $\qquad\square$

The next theorem gives a useful consequence of compactness.

Theorem 1.38. *Let X be a compact space. Suppose $\{F_\alpha\}$ is a family of closed subsets of X, any finitely many of which have nonempty intersection. Then $\bigcap F_\alpha$ is nonempty.*

Proof. For each α let $U_\alpha = F_\alpha^c$. Suppose $\bigcap F_\alpha$ is empty. Then $\bigcup U_\alpha = X$. (For any $x \in X$ there exists α such that $x \notin F_\alpha$, and hence $x \in U_\alpha$.)

Thus $\{U_\alpha\}$ is an open cover of X. By compactness, there exist indices $\alpha_1, \ldots, \alpha_n$ such that

$$U_{\alpha_1} \cup \cdots \cup U_{\alpha_n} = X.$$

But then

$$F_{\alpha_1} \cap \cdots \cap F_{\alpha_n} = \emptyset,$$

contradicting the hypothesis. We conclude that the full intersection $\bigcap F_\alpha$ must have been nonempty. $\qquad\square$

In fact, the property described in Theorem 1.38 is equivalent to compactness. We leave the reverse direction as an exercise.

Next we prove two basic results about compactness.

Proposition 1.39. *Let $f : X \to Y$ be a continuous function between two topological spaces. If X is compact then so is $f(X)$.*

Proof. Suppose X is compact and let $\mathcal{O}$ be an open cover of $f(X)$. Then $\{f^{-1}(U) : U \in \mathcal{O}\}$ is an open cover of X, and since X is compact we can find $U_1, \ldots, U_n \in \mathcal{O}$ such that

$$X = f^{-1}(U_1) \cup \cdots \cup f^{-1}(U_n).$$

It follows that $f(X) \subseteq U_1 \cup \cdots \cup U_n$, which means that $\mathcal{O}$ has a finite subcover. We conclude that $f(X)$ is compact. $\qquad\square$

Since the restriction of any continuous function to a subset equipped with the relative topology is still continuous (exercise), it follows that if $f : X \to Y$ is a continuous map between any two topological spaces then the image under f of any compact subset of X is a compact subset of Y.

Proposition 1.40. *Let X be a topological space.*

(a) If X is compact then any closed subset of X is compact.
(b) If X is Hausdorff then any compact subset of X is closed.

Proof. (a) Suppose X is compact, let $C \subseteq X$ be closed, and let $\mathcal{O}$ be a family of open subsets of X that covers C. Then $\mathcal{O} \cup \{C^c\}$ is an open cover of X, so that by compactness of X we must have

$$X = U_1 \cup \cdots \cup U_n \cup C^c$$

for some $U_1, \ldots, U_n \in \mathcal{O}$. It then follows that $C \subseteq U_1 \cup \cdots \cup U_n$, and we conclude that C is compact.

(b) Suppose X is Hausdorff and $C \subseteq X$ is compact. We claim that for every $x \in C^c$ there exists an open set U that contains x and is disjoint from C. This will establish that C^c is open, and hence that C is closed.

To prove the claim, let $x \in C^c$. For each $y \in C$, since X is Hausdorff we can find disjoint open sets U_y and V_y such that $x \in U_y$ and $y \in V_y$. The sets V_y then constitute an open cover of C, so let $V_{y_1}, \ldots, V_{y_n}$ be a finite subcover. Then $U_{y_1} \cap \cdots \cap U_{y_n}$ is an open neighborhood of x and it does not intersect $V_{y_1} \cup \cdots \cup V_{y_n}$, hence it does not intersect C. This completes the proof. $\qquad\square$

The technique used in the proof of part (b) will appear again later. In general the construction involves a compact set C, and for each $x \in C$ a pair of disjoint open sets U_x, V_x such that $x \in U_x$. Then C is covered by finitely many sets $U_{x_1}, \ldots, U_{x_n}$, and we can form the sets $U = U_{x_1} \cup \cdots \cup U_{x_n}$, which contains C, and $V = V_{y_1} \cap \cdots \cap V_{y_n}$, which is disjoint from U. We call this a *union-intersection construction*.

Combining the preceding results yields a surprising conclusion. Under a simple compactness assumption any continuous bijection is automatically a homeomorphism.

Corollary 1.41. *Let $f : X \to Y$ be a continuous bijection between topological spaces. Suppose X is compact and Y is Hausdorff. Then f is a homeomorphism.*

Proof. We must show that f^{-1} is continuous. Since $(f^{-1})^{-1} = f$, by Proposition 1.20 it will suffice to show that for any closed $C \subseteq X$ the set $f(C) \subseteq Y$ is also closed. Thus let $C \subseteq X$ be closed. By Proposition 1.40 (a), C is compact; it follows from Proposition 1.39 that $f(C)$ is compact; and it then follows from Proposition 1.40 (b) that $f(C)$ is closed. $\qquad\square$

To see that compactness is needed in this result, let Y be a Hausdorff space and let X be the same set with a strictly stronger topology. As long as Y is not discrete such a topology will always exist. Then the identity map will be a continuous bijection from X to Y whose inverse is not continuous.

Corollary 1.41 can be interpreted as saying that a compact Hausdorff topology can never be weakened without violating the Hausdorff condition. To see this, let X be a compact Hausdorff space and let Y be the same set with a weaker topology. Then the identity map is continuous from X to Y, so if Y is Hausdorff then the identity map must be a homeomorphism, i.e., the topology of Y is the same as the topology of X. In other words, any topology that is strictly weaker than a compact Hausdorff topology cannot be Hausdorff. This should explain the comment we made at the beginning of this section that in the Hausdorff setting compact spaces have as few open sets as possible.

We mention a useful consequence of Corollary 1.41.

Corollary 1.42. *Let X be a compact space and suppose there is a countable separating family of continuous functions from X into metrizable spaces. Then X is metrizable.*

Proof. Let (f_n) be a finite or infinite separating sequence of continuous functions from X into metrizable spaces X_n. For notational convenience, we can assume this sequence is infinite, filling it out with a string of constant maps if necessary. Then we can amalgamate the maps f_n to obtain a single map $f : X \to \prod X_n$ by setting

$$f(x) = (f_1(x), f_2(x), \ldots).$$

The map f is continuous by Corollary 1.24. It is also injective because the f_n are separating. Thus Corollary 1.41 yields that f is a homeomorphism between X and its image in $\prod X_n$. But the latter space is metrizable by Corollary 1.30, so any subset of it is also metrizable by Corollary 1.29, and this entails that X is metrizable. $\square$

1.6 Separation principles

The Hausdorff condition has surprisingly strong consequences in the presence of compactness. For instance, we mentioned earlier that it can be hard to tell whether quotient spaces are Hausdorff. This is essentially a matter of being able to separate distinct equivalence classes by open sets which are, in an obvious sense, compatible with the equivalence relation. Assuming compactness, we can prove an optimal result of this type. The other major result in this section, Urysohn's lemma, ensures that distinct points can be "separated" by continuous functions into $[0, 1]$.

The following enhanced version of the Hausdorff condition is basic.

Lemma 1.43. *Let X be a compact Hausdorff space and let $C, D \subseteq X$ be disjoint closed sets. Then there are disjoint open sets $U, V \subseteq X$ such that $C \subseteq U$ and $D \subseteq V$.*

Proof. Fix $x \in C$. For each $y \in D$ we can find a pair of disjoint open sets which respectively contain x and y; since D is compact, a union-intersection construction (see the comment following the proof of Proposition 1.40) then gives us a pair of disjoint open sets which respectively contain x and D. Having done this for each $x \in C$, we can use the compactness of C to perform another union-intersection construction that gives us a pair of disjoint open sets which respectively contain C and D. $\qquad \square$

We consider quotient spaces. Let X be a compact Hausdorff space. Say that an equivalence relation $\sim$ on X is *closed* if the set $\{(x, y) : x \sim y\}$ is a closed subset of $X \times X$. This is equivalent to requiring that if x is not equivalent to y then there must be a basic open set $U \times V$ in $X \times X$ which contains the pair (x, y) and avoids $\{(x, y) : x \sim y\}$. That is, if x is not equivalent to y then there must exist open sets $U, V \subseteq X$ such that $x \in U$, $y \in V$, and no element of U is equivalent to any element of V.

It is not too hard to see that if $\sim$ is not closed then $X/\sim$ cannot be Hausdorff (exercise). Conversely, in the compact Hausdorff setting, any quotient by a closed equivalence relation must be Hausdorff.

Lemma 1.44. *Let $\sim$ be a closed equivalence relation on a compact Hausdorff space X and let U be an open subset of X. Then*

$$\{x \in X : [x] \subseteq U\}$$

is an open subset of X.

Proof. Fix x such that $[x] \subseteq U$; we must find an open set U' containing x such that $[x'] \subseteq U$ for all $x' \in U'$.

The complement of U is closed and hence compact. Since $\sim$ is a closed equivalence relation, for every $y \in U^c$ there exist open sets U_y and V_y respectively containing x and y and such that no element of U_y is equivalent to any element of V_y. A union-intersection construction then produces open sets U' and V' respectively containing x and U^c such that no element of U' is equivalent to any element of V'. Since $U^c \subseteq V'$, this means that $x' \in U'$ implies $[x'] \subseteq U$, as desired. $\qquad \square$

Theorem 1.45. *Let $\sim$ be a closed equivalence relation on a compact Hausdorff space X. Then the quotient space $X/\sim$ is also compact Hausdorff.*

Proof. Compactness of $X/\sim$ follows from Proposition 1.39, since the quotient map π is a continuous surjection from X onto $X/\sim$. To see that $X/\sim$ is Hausdorff, let $[x]$ and $[y]$ be distinct elements of $X/\sim$. Then $[x]$ and $[y]$ can also be viewed as disjoint subsets of X, and the fact that $\sim$ is a closed equivalence relation implies that they are closed subsets. (X

is homeomorphic to $\{x\} \times X \subseteq X \times X$, and under this bijection $[x] \subset X$ corresponds to $\{(x, x') : x' \sim x\}$, which is closed for the relative topology on $\{x\} \times X$. Similarly for $[y]$.)

Use Lemma 1.43 to find disjoint open sets $U, V \subset X$ which respectively contain the closed sets $[x]$ and $[y]$. Then according to Lemma 1.44

$$U' = \{x' : [x'] \subseteq U\} \qquad \text{and} \qquad V' = \{y' : [y'] \subseteq V\}$$

are disjoint open sets which respectively contain $[x]$ and $[y]$. It follows that $\pi(U')$ and $\pi(V')$ are disjoint open subsets of $X/\sim$ which respectively contain the points $[x]$ and $[y]$. $\qquad \square$

In the next theorem, a recursive application of Lemma 1.43 leads to a surprising conclusion. This result is important because it implies that there are plenty of continuous scalar-valued functions on any compact Hausdorff space.

We use the following variant form of Lemma 1.43. Let X be a compact Hausdorff space and suppose $C \subseteq V \subseteq X$ with C closed and V open. Then there is an open subset $U \subseteq X$ which contains C and whose closure is contained in V. This is just a restatement of Lemma 1.43 with $V = D^c$.

Theorem 1.46. *(Urysohn's lemma) Let X be a compact Hausdorff space and let $C, D \subseteq X$ be disjoint closed subsets of X. Then there is a continuous function $f : X \to [0,1]$ such that $f|_C = 0$ and $f|_D = 1$.*

Proof. We first construct a family of open sets U_r, with r ranging over the rationals in $[0,1]$, such that (1) $C \subseteq U_0$, (2) $U_1 \subseteq D^c$, and (3) $\overline{U}_r \subseteq U_{r'}$ for all $r < r'$. To do this, set $r_1 = 1$ and $r_2 = 0$, and let (r_n) for $n \geq 3$ enumerate the rationals in $(0,1)$. Start by letting $U_{r_1} = U_1 = D^c$ and applying the variant form of Lemma 1.43 to C and U_1 to obtain an open set $U_{r_2} = U_0$ satisfying $C \subseteq U_0 \subseteq \overline{U}_0 \subseteq U_1$. At the nth stage of the construction for $n \geq 3$, consider the values $r_1, \ldots, r_{n-1}$. This is a finite, linearly ordered subset of $[0,1]$, and r_n lies between two adjacent values r_i and r_j in this set. Use Lemma 1.43 to find an open set U_{r_n} such that

$$\overline{U}_{r_i} \subseteq U_{r_n} \subseteq \overline{U}_{r_n} \subseteq U_{r_j}.$$

This procedure generates a family of open sets with the desired properties.

Now for each $t \in [0,1]$ let

$$A_t = \left(\bigcap_{r>t} U_r\right) \setminus \left(\bigcup_{r<t} U_r\right),$$

taking the intersection to be all of X when $t = 1$ and the union to be null when $t = 0$. A moment's thought shows that the sets A_t are disjoint and their union equals X. So we can define $f : X \to [0, 1]$ by letting $f(x) = t$ for all $x \in A_t$. It is clear that $f|_C = 0$ and $f|_D = 1$.

Finally, to see that f is continuous, let $a \in (0, 1)$. Then

$$f^{-1}([0, a)) = \bigcup_{t<a} A_t = \bigcup_{r<a} U_r$$

and

$$f^{-1}((a, 1]) = \bigcup_{t>a} A_t = \bigcup_{r>a} (\overline{U}_r)^c$$

are both open, and as the intervals of the form $[0, a)$ and $(a, 1]$ generate the topology on $[0, 1]$, it follows that f is continuous. $\qquad\square$

We can use Urysohn's lemma to characterize which compact spaces are metrizable.

Corollary 1.47. *Let X be a compact Hausdorff space. Then X is metrizable iff X is second countable.*

Proof. If there is a metric on X that gives rise to its topology, then for each $n \in \mathbf{N}$ we can extract a finite subcover of X from $\{\mathrm{ball}_{1/n}(x) : x \in X\}$. The centers of the balls in the resulting subcovers will then constitute a countable subset A of X such that for each $n \in \mathbf{N}$, every element of X is within $\frac{1}{n}$ of some element of A. So X is separable, and hence it is second countable by Proposition 1.32. Conversely, suppose X is second countable and let $\mathcal{B}$ be a countable base for its topology. For any $U, V \in \mathcal{B}$ with $\overline{U}$ and $\overline{V}$ disjoint, use Urysohn's lemma to find a continuous function from X into $[0, 1]$ which is 0 on U and 1 on V. This gives us a countable family of continuous functions $f_{U,V} : X \to [0, 1]$. Moreover, if $x, y \in X$ are distinct then using Lemma 1.43 we can find open sets U and V respectively containing x and y whose closures are disjoint; but then there must be basic open sets with the same properties, and this shows that $f_{U,V}(x) \neq f_{U,V}(y)$ for some U, V. We conclude from Corollary 1.42 that X is metrizable. $\qquad\square$

1.7   Local compactness

Local compactness is a weakened version of the notion of compactness that encompasses a much larger class of examples.

Definition 1.48. A topological space X is *locally compact* if every element of X has a compact neighborhood.

Obviously, any compact space is locally compact since the entire space is a neighborhood of each of its points. Any discrete space is also locally compact. The real line is another example: for any $a, b \in \mathbf{R}$, $a < b$, the closed interval $[a, b]$ is homeomorphic to $[0, 1]$ and hence is compact by Theorem 1.37, and it follows that every element of $\mathbf{R}$ has a compact neighborhood.

The relationship between compactness and local compactness is clarified in the next result. It states that the locally compact Hausdorff spaces are (up to homeomorphism) precisely those topological spaces which can be obtained by removing a single point from a compact Hausdorff space. To a large extent this fact reduces the theory of locally compact Hausdorff spaces to the theory of compact Hausdorff spaces.

Definition 1.49. Let X be a locally compact Hausdorff space. The *one-point compactification* of X is the set $X^* = X \cup \{\infty\}$ equipped with the topology according to which $U \subseteq X^*$ is open if either (1) U is an open subset of X or (2) $\infty \in U$ and $X^* \setminus U$ is a compact subset of X.

Here ∞ is just a symbol for some new point which is being added to X. This notation is motivated by examples like $\mathbf{R}$ where one intuitively thinks of adding a point "at infinity".

If X is already compact then the singleton set $\{\infty\}$ will be open, so we have just added a new point which is, in this sense, isolated. This version of the construction is not so important.

It is implicit in Definition 1.49 that the prescription given there does define a topology on X^*. We leave this verification as an exercise.

Proposition 1.50. *Let X be a Hausdorff space.*

(a) If X is compact and $x \in X$ then $X \setminus \{x\}$ is locally compact.
(b) If X is locally compact then X^* is a compact Hausdorff space, and the relative topology on $X \subset X^*$ is the original topology on X.

Proof. (a) Suppose X is compact and let $x \in X$. Fix $y \in X \setminus \{x\}$; we must find a compact neighborhood of y. By the Hausdorff property there exists an open neighborhood V of y whose closure $\overline{V}$ avoids x. As $\overline{V}$ is compact by Proposition 1.40 (a), we see that y has a compact neighborhood in $X \setminus \{x\}$, as desired.

(b) Suppose X is locally compact. Then X^* is compact because any open cover $\mathcal{O}$ of X^* must include a set that contains ∞, and the complement of that set will be a subset of X which is compact and hence can be covered by a finite subfamily of $\mathcal{O}$. It is Hausdorff because any element x of X has a compact neighborhood C in X; since C is a neighborhood of x it contains an open neighborhood of x, and since it is compact its complement in X^* is an open neighborhood of ∞. So any point of X can be separated from ∞ by open sets. Any two distinct elements of X can be separated by open sets in X^* the same way this was done in X.

It is clear that the restriction of the topology on X^* to X includes every set that was originally open in X. We do not add any new open sets because every compact subset of X is closed, which implies that the intersection of any open neighborhood of ∞ with X is an open subset of X. $\qquad\square$

Since we are most interested in metrizable spaces, we should ask whether the one-point compactification of a metrizable locally compact space is always metrizable. The answer is no. For instance, any uncountable set with the discrete topology is metrizable and locally compact, but its one-point compactification is not metrizable. However, assuming second countability we get a positive answer.

Proposition 1.51. *Let X be a second countable locally compact Hausdorff space. Then X^* is metrizable.*

Proof. By Corollary 1.47 we just need to show that X^* is second countable. Let (U_n) be a countable base for X. By local compactness we can assume that every U_n has compact closure. (Discard every set whose closure is not compact; the remaining sets must still constitute a base.)

For each $n \in \mathbf{N}$ let V_n be the complement of $\overline{U}_1 \cup \cdots \cup \overline{U}_n$ in X^*. We will show that the sets U_n and V_n together constitute a base for X^*.

It will suffice to show that any open subset of X^* which contains ∞ must contain some V_n. So let U be an open subset of X^* which contains ∞. Then U^c must be compact; since $U^c \subseteq X = \bigcup_{i=1}^{\infty} U_i$, it follows from compactness that $U^c \subseteq \bigcup_{i=1}^{n} U_i$ for some n. But this means that $V_n \subseteq U$, which is what we needed to show. $\qquad\square$

It follows that any second countable locally compact Hausdorff space is metrizable.

Second countability is necessary in Proposition 1.51 because according to Corollary 1.47 any compact metrizable space is second countable, and if

X^* is second countable then X must also be second countable by Proposition 1.34.

1.8 Sequential convergence

One of the main advantages of working with metrizable spaces is that the topology is determined by the behavior of sequences. This can be quite useful, both as a technical tool and as an aid to intuition.

Definition 1.52. Let X be a topological space and let (x_n) be a sequence of points in X. We say that (x_n) *converges to* $x \in X$ if it eventually lies in every neighborhood of x, that is, for any neighborhood U of x, there exists $N \in \mathbf{N}$ such that

$$n \geq N \quad \Rightarrow \quad x_n \in U.$$

We call x the *limit* of the sequence (x_n) and write both $x_n \to x$ and $x = \lim x_n$.

In a metric space, every neighborhood of a point contains a ball about that point, and every open ball about a point is a neighborhood of that point. So a sequence converges to x if and only if it eventually belongs to any ball about x. This yields the following characterization.

Example 1.53. Let X be a metric space, let $x \in X$, and let (x_n) be a sequence in X. Then $x_n \to x$ iff for every $r > 0$ there exists $N \in \mathbf{N}$ such that

$$n \geq N \quad \Rightarrow \quad d(x_n, x) < r.$$

We also note that if the topology on X is generated by a family $\mathcal{S}$, then we can verify that $x_n \to x$ by checking only that x_n eventually belongs to every set in $\mathcal{S}$ that contains x. This is because for any neighborhood U of x there exist finitely many sets $U_1, \ldots, U_k \in \mathcal{S}$ such that $x \in U_1 \cap \cdots \cap U_k \subseteq U$, and if (x_n) eventually belongs to each U_i then it must eventually belong to their intersection, and hence to U.

Say that a subset C of a topological space X is *sequentially closed* if whenever a sequence of points in C converges to a point in X, the limit point belongs to C. Say that a function $f : X \to Y$ from one topological space to another *preserves sequential limits* if $x_n \to x$ in X implies $f(x_n) \to f(x)$ in Y.

Proposition 1.54. *Let X and Y be topological spaces and assume X is metrizable.*

(a) A subset $C \subseteq X$ is closed iff it is sequentially closed.
(b) A function $f : X \to Y$ is continuous iff it preserves sequential limits.

Proof. (a) ($\Rightarrow$) Let $C \subseteq X$ be closed and let (x_n) be a sequence in C which converges to some point $x \in X$. Suppose $x \notin C$. Since C is closed, its complement is open, and hence the sequence (x_n) eventually lies in C^c. But this contradicts the hypothesis that (x_n) lies entirely in C. We conclude that x must belong to C. This shows that C is sequentially closed.

($\Leftarrow$) Fix a metric that gives rise to the topology on X and suppose $C \subseteq X$ is not closed. Then its complement is not open, so we can find $x \in C^c$ such that every open ball about x intersects C. For each $n \in \mathbf{N}$ choose $x_n \in \mathrm{ball}_{1/n}(x) \cap C$. Then (x_n) is a sequence in C which converges to $x \notin C$, so C is not sequentially closed.

(b) ($\Rightarrow$) Let $f : X \to Y$ be a continuous function and suppose $x_n \to x$ in X. Let $U \subseteq Y$ be a neighborhood of $f(x)$. Then $f^{-1}(U)$ is a neighborhood of x, so there must exist $N \in \mathbf{N}$ such that $x_n \in f^{-1}(U)$ for all $n \geq N$. This implies that $f(x_n) \in U$ for all $n \geq N$. We conclude that $f(x_n) \to f(x)$ in Y, so that f preserves sequential limits.

($\Leftarrow$) Suppose $f : X \to Y$ is not continuous. Then there is a closed subset C of Y whose inverse image under f is not closed in X. By part (a), there must exist a sequence (x_n) in $f^{-1}(C)$ that converges to a point x in its complement. Then $(f(x_n))$ is contained in C but $f(x)$ is not, so we cannot have $f(x_n) \to f(x)$. This shows that f does not preserve sequential limits. $\square$

The hypothesis that X is metrizable was not actually used in the forward direction of either part of this proposition, and in the reverse direction all we really used was that X has a countable neighborhood base at each point, that is, every $x \in X$ has a sequence of neighborhoods (U_n) such that every neighborhood of x contains some U_n. For instance, this would also follow from the assumption that X is second countable.

Weak topologies can also be understood in terms of sequential convergence. The following observation is useful here. Let (x_n) be a sequence in a topological space X and let $x \in X$. Then we claim that x_n converges to x if and only if the map taking n to x_n and ∞ to x is continuous from the one-point compactification $\mathbf{N}^*$ of $\mathbf{N}$ into X. The verification of this claim is routine.

Proposition 1.55. *Let X be a set equipped with the weak topology determined by a family of maps $f_\alpha : X \to X_\alpha$ into topological spaces X_α. Then a sequence (x_n) in X converges to $x \in X$ iff $f_\alpha(x_n) \to f_\alpha(x)$ for all α.*

In light of the observation made just above, this proposition is an immediate consequence of Proposition 1.22.

This yields an interpretation of product topologies which may be more intuitive than their characterization in terms of open sets.

Corollary 1.56. *Let (X_n) be a finite or infinite sequence of topological spaces. Then a sequence in $\prod X_n$ converges to a point iff it converges to that point on each coordinate.*

Compactness can also be characterized sequentially, but this equivalence uses metrizability in a more substantial way than the preceding results did. Say that a metric space X is *totally bounded* if for any $\epsilon > 0$ there is a finite set of points $x_1, \ldots, x_n \in X$ such that

$$X = \mathrm{ball}_\epsilon(x_1) \cup \cdots \cup \mathrm{ball}_\epsilon(x_n).$$

That is, every element of X is within ϵ of some x_i.

Theorem 1.57. *Let X be a metric space. Then X is compact iff X is complete and totally bounded iff every sequence in X has a convergent subsequence.*

Proof. Assume X is compact; we show that it is complete and totally bounded. Let $\overline{X}$ be the completion of X and suppose $X \neq \overline{X}$. Then there exists $x \in \overline{X} \setminus X$, and the sets $U_n = \{y \in X : d(x,y) > \frac{1}{n}\}$ constitute an open cover of X with no finite subcover, contradicting compactness. So X must be complete. Also, for any $\epsilon > 0$ we can extract a finite subcover from $\{\mathrm{ball}_\epsilon(x) : x \in X\}$, which shows that X must be totally bounded.

Now suppose X is complete and totally bounded, and let (x_n) be any sequence in X. By total boundedness we can find a finite family of balls of radius 1 which cover X. It follows that there exists a point y_1 such that the set

$$A_1 = \{n \in \mathbf{N} : x_n \in \mathrm{ball}_1(y_1)\}$$

is infinite. Let n_1 be the smallest element of A_1. By total boundedness again, there exists a point y_2 such that the set

$$A_2 = \{n \in A_1 : x_n \in \mathrm{ball}_{1/2}(y_2)\}$$

is infinite; let n_2 be the smallest element of A_2 that is greater than n_1. Continuing this process yields a Cauchy subsequence (x_{n_k}) of the sequence (x_n), and this subsequence then converges by completeness. Thus, we have shown that every sequence has a convergent subsequence.

Finally, suppose every sequence has a convergent subsequence and let $\mathcal{O}$ be an open cover of X. We claim that there exists $r > 0$ such that for every $x \in X$ the set $\text{ball}_r(x)$ is contained in some $U \in \mathcal{O}$. If this claim fails, then for each n we can find a point x_n such that $\text{ball}_{1/n}(x_n)$ is not contained in any $U \in \mathcal{O}$. Let x be the limit of a convergent subsequence of (x_n); then x belongs to some $U \in \mathcal{O}$ and hence some ball about x is contained in this U, which easily yields a contradiction. This proves the claim. Now let y_1 be any point in X, and successively find points y_2, y_3, ... such that the distance from y_n to each y_i with $i < n$ is at least r. If this process were to continue indefinitely then it would create a sequence with no convergent subsequence, which is impossible. Therefore it must terminate, which means that the finitely many sets $\text{ball}_r(y_i)$ cover X, and by the claim it then follows that $\mathcal{O}$ has a finite subcover. Thus X is compact. $\qquad\square$

Corollary 1.58. *A metrizable space is compact if and only if every sequence in it has a convergent subsequence.*

The following simple proof of Tychonoff's product theorem, which covers all the cases of interest to us, illustrates the power of sequential convergence.

Theorem 1.59. *(Tychonoff product theorem) The product of a countable family of compact metrizable spaces is compact.*

Proof. Let (X_n) be a finite or infinite sequence of compact metrizable spaces. For notational convenience, we can assume the sequence is infinite because a finite sequence can be filled out with an infinite string of one-element spaces without (up to homeomorphism) affecting the product.

We know that $\prod X_n$ is metrizable by Corollary 1.30. According to Corollary 1.56, a sequence converges in $\prod X_n$ if and only if it converges on each coordinate, and by Corollary 1.58, to show that $\prod X_n$ is compact it will suffice to show that every sequence has a convergent subsequence. To do this, let $(\mathbf{x}^k)$ be a sequence in $\prod X_n$. Find a subsequence $(\mathbf{x}^{k_i})$ which converges on the first coordinate, say $x_1^{k_i} \to x_1$. Let $\mathbf{y}^1 = \mathbf{x}^{k_1}$. Then find a subsequence $(\mathbf{x}^{k_i'})$ of $(\mathbf{x}^{k_i})$ which converges on the second coordinate, say $x_2^{k_i'} \to x_2$, and let $\mathbf{y}^2 = \mathbf{x}^{k_2'}$. Proceed in this way. The result is a sequence $(\mathbf{y}^j)$ which is a subsequence of $(\mathbf{x}^k)$ and which converges to the point $\mathbf{x} = (x_n)$. So we have shown that every sequence has a convergent subsequence. □

This result allows us to neatly characterize the compact subsets of $\mathbf{R}^n$.

Corollary 1.60. *(Heine-Borel theorem) A subset of $\mathbf{R}^n$ is compact iff it is closed and bounded.*

Proof. Any compact subset of $\mathbf{R}^n$ must be closed and bounded by Theorem 1.57. Conversely, suppose $C \subset \mathbf{R}^n$ is closed and bounded. Then for some $M > 0$ it is a closed subset of the cube $[-M, M]^n$. Therefore, C is compact by Theorem 1.37, Theorem 1.59, and Proposition 1.40 (a). □

It follows that $\mathbf{R}^n$ is locally compact.

1.9 Exercises

1.1. Prove that the set of points in $\mathbf{R}^n$ whose coordinates are all rational is countable.

1.2. Prove that the set of all sequences of rational numbers in which only finitely many terms are nonzero is countable.

1.3. Prove that the set of all sequences of natural numbers is uncountable.

1.4. Prove that the set of all subsets of $\mathbf{N}$ is uncountable.

1.5. Find an uncountable family of subsets of $\mathbf{N}$ any two of which have finite intersection.

1.6. Verify that the metric topology associated to any metric space really is a topology.

1.7. Prove that the discrete topology on any set is the metric topology for some metric on that set.

1.8. Is every Hausdorff topology on a finite set a discrete topology?

1.9. Let $\mathcal{B}$ be a family of subsets of a set X. Prove that $\mathcal{B}$ is a base for a topology iff (1) every element of X belongs to some set in $\mathcal{B}$ and (2) if $U_1, U_2 \in \mathcal{B}$ and $x \in U_1 \cap U_2$ then there exists $U \in \mathcal{B}$ such that $x \in U \subseteq U_1 \cap U_2$.

1.10. Prove that the discrete topology on any set X is the weak topology determined by a family of functions from X into a two-element discrete space.

1.11. Prove that the metric topology on any metric space X is the weak topology determined by a family of functions from X into $\mathbf{R}$.

1.12. Let $f : X \to Y$ be a continuous function between topological spaces and let $A \subseteq X$. Prove that the restriction of f to A is continuous with respect to the relative topology on A.

1.13. Prove that the real line is homeomorphic to $(0,1)$ but not to $[0,1)$ (equipping both of the latter sets with the relative topology inherited from **R**).

1.14. Prove Proposition 1.19.

1.15. A topological space X is *connected* if $\emptyset$ and X are the only subsets which are simultaneously open and closed. Prove that X is connected iff there is no continuous function from X onto a two-element discrete space.

1.16. Prove that the metric topology on any separable metric space X is the weak topology determined by a countable family of functions from X into **R**.

1.17. Find an example of a Hausdorff space that is not metrizable.

1.18. Let X be a topological space with the property that if $\{F_\alpha\}$ is a family of closed subsets of X any finitely many of which have nonempty intersection, then $\bigcap F_\alpha$ is nonempty. Prove that X is compact.

1.19. Let $f : X \to \mathbf{R}$ be a continuous function from a compact space into **R**. Prove that f is bounded and that it attains maximum and minimum values.

1.20. Let X be a topological space and let $\sim$ be an equivalence relation on X which is not closed. Show that $X/\sim$ is not Hausdorff.

1.21. Find an example of a closed equivalence relation on a separable metric space such that the quotient space is not metrizable.

1.22. Prove that Definition 1.49 does define a topology on X^*.

1.23. Let X be a locally compact metrizable space. Prove that every open subset of X is a locally compact metrizable space.

1.24. Let X be a compact Hausdorff space and let $x \in X$. Prove that X is homeomorphic to the one-point compactification of $X \setminus \{x\}$.

1.25. Let A be a subset of a metrizable space X. Prove that $\overline{A}$ is precisely the set of points in X which are limits of sequences in A.

The *Cantor set*

$$\mathbf{K} = [0,1] \setminus \left(\left(\frac{1}{3}, \frac{2}{3} \right) \cup \left(\frac{1}{9}, \frac{2}{9} \right) \cup \left(\frac{7}{9}, \frac{8}{9} \right) \cup \cdots \right)$$

is what is left of the unit interval after its open middle third is removed, the open middle thirds of the two remaining intervals are removed, etc.

1.26. Prove that $\mathbf{K}$ is uncountable. (Draw a picture of the proof of Theorem 1.6.)

1.27. Prove that there is no continuous map from $\mathbf{K}$ onto $(0,1)$.

1.28. Prove that there is a continuous map from **K** onto $[0, 1]$.

1.29. Prove that **K** is homeomorphic to the product of a sequence of two-element discrete spaces.

1.30. Prove or disprove: for any $x, y \in \mathbf{K}$ there is a homeomorphism of **K** with itself that takes x to y.

Chapter 2

Measure and Integration

2.1 Measurable spaces and functions

A measure on a set is a function which assigns numerical values to subsets of the set. The crucial feature is an additivity condition which requires that the measure of a disjoint union equals the sum of the measures of the component sets. A basic motivating example is the notion of volume for subsets of $\mathbf{R}^n$, and a good intuition in general is to think of a measure as arising from a distribution of mass, with the measure of any subset being the total mass it contains. (Volume measure in $\mathbf{R}^n$ corresponds to the case of a uniform distribution of mass with unit density.) For measures which take both positive and negative values, one can imagine a distribution of positive and negative charge.

Already in the prototypical case of volume in $\mathbf{R}^n$ we find that it is unreasonable to demand that measures be defined for all subsets. There are pathological subsets of $\mathbf{R}^n$ for which we have no meaningful notion of volume, indeed, sets for which any attempt to assign a volume has paradoxical consequences; see Example 2.21. We handle this issue by asking measures to be defined not for all subsets, but merely for a family of subsets with good stability properties. The relevant definition is the following.

Definition 2.1. A *σ-algebra of subsets* of a set X is a family Ω of subsets of X which satisfies

 (i) $\emptyset, X \in \Omega$;
 (ii) if $A \in \Omega$ then $A^c \in \Omega$;
 (iii) if (A_n) is a sequence of sets in Ω then $\bigcup A_n, \bigcap A_n \in \Omega$.

In words: Ω contains $\emptyset$ and X, and it is stable under complementation, countable unions, and countable intersections.

We call a set equipped with a σ-algebra of subsets a *measurable space*. A subset is *measurable* if it belongs to the σ-algebra.

A σ-algebra is the background structure that is needed to support a measure. As we will see, stability under countable, not merely finite, set theoretic operations is important because it enables us to make approximation arguments. To see why this kind of stability is reasonable, let (A_n) be any sequence of sets in $\mathbf{R}^n$ whose volume is well-defined and consider the finite unions $B_n = A_1 \cup \cdots \cup A_n$. Then the volume of B_n should increase with n, and it is not hard to convince oneself that $\bigcup A_n = \bigcup B_n$ ought to have a well-defined volume which is the limit of the volumes of the B_n. If the volume of B_n goes to infinity, then surely their union has infinite volume, and if it increases to a finite value then the volume in the "tail" $(\bigcup_{n=1}^{\infty} B_n) \setminus B_N$ should be going to zero as $N \to \infty$.

We will develop the concept of a σ-algebra in this section and then present the axioms for a measure in the next section. Before proceeding, we should note that the conditions in Definition 2.1 are somewhat redundant. For instance, stability under countable intersections does not have to be assumed separately. It follows from stability under complements and countable unions because the intersection of a sequence of sets can also be described as the complement of the union of their complements. With a little more work we can weaken the conditions still further:

Proposition 2.2. *Let X be a set and let Ω be a nonempty family of subsets of X. Suppose that Ω is stable under complements, finite unions, and countable disjoint unions. Then Ω is a σ-algebra.*

Proof.   Since Ω is nonempty, it contains some set A, and hence it contains $X = A \cup A^c$ and $\emptyset = X^c$. To verify stability under countable unions, let (A_n) be a sequence of sets in Ω. Define $B_1 = A_1$ and $B_n = A_n \setminus (A_1 \cup \cdots \cup A_{n-1})$ for $n \geq 2$. Since B_n can also be expressed as the complement of $A_1 \cup \cdots \cup A_{n-1} \cup A_n^c$, it belongs to Ω. The sequence (B_n) is disjoint, so by hypothesis its union belongs to Ω; but $\bigcup A_n = \bigcup B_n$, so we conclude that $\bigcup A_n \in \Omega$. This shows that Ω is stable under arbitrary countable unions. Stability under countable intersections now follows by the remark made just before this proposition. We conclude that Ω is a σ-algebra. $\square$

The point of this proposition is merely to streamline the task of verifying that something is a σ-algebra. There are other ways to minimize the conditions that need to be checked, but this is the best version for our purposes.

It would be nice to present some examples at this point. The simplest way to construct a σ-algebra on a set is to declare that all subsets are measurable, just as we did for discrete topological spaces. This is actually an important special case, and we will say more about it in the next section. Unfortunately, it is not so easy to give concrete examples of σ-algebras that are significantly more general than this. Nonconstructive examples can be readily produced, but these are not very instructive, for obvious reasons. So we defer the description of more serious examples to Sections 2.4 and 4.5.

We will be interested in functions from measure spaces into topological spaces, particularly into the scalar field. We want such functions to be compatible with the topological and measurable structure in a certain way.

Definition 2.3. Let X be a measurable space and Y a topological space. A function $f : X \to Y$ is *measurable* if $f^{-1}(U)$ is measurable in X for every open subset U of Y.

This condition might seem too strong, since topologies are stable under arbitrary unions and σ-algebras are only stable under countable unions. But recall that if a topological space is second countable then every open set can be expressed as a union of countably many basic open sets (since there are only countably many basic open sets in total). So the discrepancy is not as extreme as it looks at first.

It is probably not too surprising to learn that sums and products of measurable functions are measurable. Indeed, there is a general technique for establishing this kind of result (see the proof of Theorem 2.6 (c) below). More interestingly, pointwise limits of sequences of measurable functions are also measurable. Before proving this result we need two lemmas. The first one relies on the following simple observation. Let X be a measurable space, Y a set, and $f : X \to Y$ a function. Then

$$\{A \subseteq Y : f^{-1}(A) \subseteq X \text{ is measurable}\}$$

is a σ-algebra of subsets of Y. This follows from the fact that set theoretic operations commute with inverse images (see the comment just after Definition 1.18).

Lemma 2.4. *Let X be a measurable space and let $f : X \to \mathbf{R}$ be a function. Then f is measurable iff $f^{-1}((a, \infty))$ is a measurable subset of X for all $a \in \mathbf{R}$.*

Proof.   Since the intervals (a, ∞) are all open, the forward direction is immediate. For the reverse direction, suppose $f^{-1}((a, \infty))$ is a measurable subset of X for all $a \in \mathbf{R}$ and let Ω be the family of subsets A of Y with the property that $f^{-1}(A)$ is measurable in X. By the observation made just above, Ω is a σ-algebra, and we know that it contains the intervals (a, ∞); we must show that it contains every open subset.

First, since $(a, \infty) \in \Omega$ for all $a \in \mathbf{R}$, by taking complements we get that $(-\infty, a] \in \Omega$ for all $a \in \mathbf{R}$. Then taking a countable union of the intervals $(-\infty, b - \frac{1}{n}]$ shows that $(-\infty, b) \in \Omega$, for all $b \in \mathbf{R}$. Next, we have $(a, b) = (-\infty, b) \cap (a, \infty) \in \Omega$ for all $a < b$. Finally, since any open subset of $\mathbf{R}$ is a countable union of open intervals, it follows that Ω contains every open subset, as desired. $\square$

Lemma 2.5. *Let X be a measurable space and let $f : X \to \mathbf{R}^2$ be a function. Then f is measurable iff both of its components are measurable.*

Proof.   Let f_1 and f_2 be the components of f, so that $f(x) = (f_1(x), f_2(x))$ for all $x \in X$. Also let $\pi_1, \pi_2 : \mathbf{R}^2 \to \mathbf{R}$ be the coordinate projections. If f is measurable then $f_1 = \pi_1 \circ f$ and $f_2 = \pi_2 \circ f$ are measurable because π_1 and π_2 are continuous. Conversely, suppose f_1 and f_2 are measurable. If $U, V \subseteq \mathbf{R}$ are open then

$$f^{-1}(U \times V) = f_1^{-1}(U) \cap f_2^{-1}(V)$$

is a measurable subset of X. As every open subset of $\mathbf{R}^2$ is a countable union of open sets of the form $U \times V$, it follows that the inverse image of any open set is measurable. $\square$

We are interested in both real and complex scalars. Since much of the theory works the same way in the two cases, it will be convenient to have a single symbol $\mathbf{F}$ which can denote either $\mathbf{R}$ or $\mathbf{C}$.

Theorem 2.6. *Let X be a measurable space.*

(a) *Any constant function from X into $\mathbf{F}$ is measurable.*

(b) *The complex conjugate of any measurable function from X into $\mathbf{C}$ is measurable. A function from X into $\mathbf{C}$ is measurable iff its real and imaginary parts are both measurable.*

(c) *The sum and product of two measurable functions from X into $\mathbf{F}$ are measurable.*

(d) *Let (f_n) be a pointwise bounded sequence of measurable functions from X into $\mathbf{R}$. Then $\sup f_n$ and $\inf f_n$, both taken pointwise, are measurable.*

(e) Let (f_n) be a sequence of measurable functions from X into $\mathbf{F}$. If the pointwise limit $\lim f_n$ exists then it is measurable.

Proof. (a) If $f : X \to \mathbf{F}$ is a constant function then for any subset A of $\mathbf{F}$ we have $f^{-1}(A) = \emptyset$ or X. Since both $\emptyset$ and X must be measurable subsets of X, it follows that f is measurable.

(b) Define $h : \mathbf{C} \to \mathbf{C}$ by $h(z) = \bar{z}$. Then measurability of f implies measurability of $\bar{f} = h \circ f$ because h is continuous. The second statement follows immediately from Lemma 2.5 by identifying $\mathbf{C}$ with $\mathbf{R}^2$.

(c) We can reduce to the real case by separating into real and imaginary parts. So let $f_1, f_2 : X \to \mathbf{R}$ be measurable and define $h, k : \mathbf{R}^2 \to \mathbf{R}$ by $h(s,t) = s + t$ and $k(s,t) = st$. Then h and k are both continuous, and the function $f : X \to \mathbf{R}^2$ whose components are f_1 and f_2 is measurable by Lemma 2.5. It follows that $f_1 + f_2 = h \circ f$ and $f_1 f_2 = k \circ f$ are measurable functions.

(d) Let $g = \sup f_n$ be the pointwise supremum of the f_n. For each $a \in \mathbf{R}$ we have

$$g^{-1}((a, \infty)) = \bigcup_{n=1}^{\infty} f_n^{-1}((a, \infty)),$$

so each of these sets is measurable in X. It follows from the lemma that g is measurable. Measurability of $\inf f_n$ is proven similarly.

(e) Reduce to the case of real scalars using part (b). Suppose the pointwise limit exists. Then the sequence is pointwise bounded and

$$\lim f_n = \limsup f_n = \inf_m \sup_{n \geq m} f_n$$

is measurable by a double application of part (d). $\qquad\square$

2.2 Positive measures

We define the notion of a measure.

Definition 2.7. Let X be a measurable space with σ-algebra Ω. A *measure* on X is a function μ from Ω into $[0, \infty]$ such that $\mu(\emptyset) = 0$ and $\mu \left(\bigcup A_n \right) = \sum \mu(A_n)$ whenever (A_n) is a sequence of disjoint measurable sets.

A *measure space* is a measurable space equipped with a measure, i.e., a set equipped with both a σ-algebra of subsets and a measure defined on that σ-algebra. It is *finite* if $\mu(X) < \infty$ and *σ-finite* if there exists a sequence (A_n) of measurable sets such that $X = \bigcup A_n$ and $\mu(A_n) < \infty$ for all n.

We may also use the term "positive measure" to avoid confusion with measures which can take negative or complex values, which we will discuss later.

Since σ-algebras enjoy such strong closure conditions, they are very comprehensive, to the extent that in typical cases one cannot even prove that nonmeasurable sets exist without using the axiom of choice. For this reason one often suppresses explicit mention of the σ-algebra and assumes that all subsets under discussion are measurable. Thus a measure space with underlying set X, σ-algebra Ω, and measure μ is usually identified as (X, μ).

The condition $\mu(\bigcup A_n) = \sum \mu(A_n)$ for sequences of disjoint sets is called *countable additivity*. The analogous condition for finite families of disjoint sets is called *finite additivity*. Finite additivity follows from countable additivity because any finite sequence can be filled out with a string of empty sets, and $\mu(\emptyset) = 0$.

We promised to say more about the σ-algebra consisting of all subsets of a set. Here is the standard measure defined on that σ-algebra:

Example 2.8. Let X be any set equipped with the σ-algebra consisting of all of its subsets. For $A \subseteq X$ define $\mu_{\text{count}}(A)$ to be the number of elements of A if A is finite, and ∞ if A is infinite. This is *counting measure* on X.

This class of examples can be generalized by means of "weight" functions. For instance, let w be any function from $\mathbf{N}$ into $[0, \infty]$ and for $A \subseteq \mathbf{N}$ define $\mu(A)$ to be $\sum_{n \in A} w(n)$. It is not hard to check that this defines a measure. But in order to produce any substantially different kind of example — any example in which mass is spread out continuously rather than being concentrated at discrete points — we are going to need some machinery. We will develop this machinery in the next section.

The basic properties of measures are collected in the following theorem. Property (a) is called "monotonicity" and property (b) is called "subadditivity." Note that in part (b) the sets (A_n) are not required to be disjoint, so there could be double counting on the right, which accounts for the inequality.

Theorem 2.9. *Let (X, μ) be a measure space, let A and B be measurable subsets of X, and let (A_n) be a sequence of measurable subsets of X.*

(a) *If $A \subseteq B$ then $\mu(A) \leq \mu(B)$.*
(b) *We have $\mu(\bigcup_{n=1}^{\infty} A_n) \leq \sum_{n=1}^{\infty} \mu(A_n)$.*
(c) *If $A_1 \subseteq A_2 \subseteq \cdots$ then $\mu(\bigcup_{n=1}^{\infty} A_n) = \lim \mu(A_n)$.*

(d) If $A_1 \supseteq A_2 \supseteq \cdots$ and $\mu(A_1) < \infty$, then $\mu(\bigcap_{n=1}^{\infty} A_n) = \lim \mu(A_n)$.

Proof. (a) If $A \subseteq B$ then $\mu(B) = \mu(B \setminus A) + \mu(A) \geq \mu(A)$.

(b) Let $B_1 = A_1$ and for $n > 1$ let $B_n = A_n \setminus (A_1 \cup \cdots \cup A_{n-1})$. Then

$$\mu\left(\bigcup A_n\right) = \mu\left(\bigcup B_n\right) = \sum \mu(B_n) \leq \sum \mu(A_n),$$

using part (a) and the fact that $B_n \subseteq A_n$.

(c) Assuming (A_n) is an increasing sequence of measurable sets, let $B_1 = A_1$ and for $n > 1$ define $B_n = A_n \setminus A_{n-1}$. Then

$$\mu\left(\bigcup A_n\right) = \mu\left(\bigcup B_n\right) = \sum \mu(B_n).$$

But $\sum_{n=1}^{\infty} \mu(B_n) = \lim_{n \to \infty} \sum_{k=1}^{n} \mu(B_k) = \lim_{n \to \infty} \mu(A_n)$, which yields the desired result.

(d) Assume (A_n) is a decreasing sequence of measurable sets and $\mu(A_1) < \infty$. We deduce the desired result by taking complements in A_1 and invoking part (c). In detail, for all n define $B_n = A_1 \setminus A_n$; applying (c) to the sequence (B_n), we obtain $\mu(\bigcup B_n) = \lim \mu(B_n)$. Thus

$$\mu\left(\bigcap A_n\right) = \mu(A_1) - \mu\left(\bigcup B_n\right) = \mu(A_1) - \lim \mu(B_n) = \lim \mu(A_n),$$

as desired. $\qquad\square$

The condition $\mu(A_1) < \infty$ in part (d) is necessary. For example, consider $\mathbf{N}$ with counting measure and let $A_n = \{n, n+1, \ldots\}$. Then $\mu_{\text{count}}(A_n) = \infty$ for all n but $\mu_{\text{count}}(\bigcap A_n) = \mu_{\text{count}}(\emptyset) = 0$.

Parts (c) and (d) are typical continuity properties that follow from countable additivity. If we only assumed finite additivity we would not be able to prove results like these.

We conclude this section with a discussion of null sets.

Definition 2.10. Let (X, μ) be a measure space. A measurable subset A of X is called a *null set* if $\mu(A) = 0$. A statement $P(x)$ is true for *almost every x* (abbreviated *a.e. x*) if there is a null set A such that $P(x)$ is true for every x not belonging to A. The measure μ is *complete* if every subset of every null set is measurable.

By part (a) of Theorem 2.9, any measurable subset of a set of measure zero must also have measure zero. However, it is possible that measure zero sets could have nonmeasurable subsets (explaining our slightly cautious formulation of the definition of "almost every"). This can always be remedied by a simple construction.

Proposition 2.11. *Let (X, μ) be a measure space with σ-algebra Ω. Define $\overline{\Omega}$ to consist of all sets of the form $A \cup B$ such that $A \in \Omega$ and B is contained in a null set, and define $\overline{\mu}(A \cup B) = \mu(A)$ for all such A and B. Then $(X, \overline{\mu})$ is a complete measure space.*

Proof.   First we check that $\overline{\Omega}$ is a σ-algebra. Let $(A_n \cup B_n)$ be a sequence of sets in $\overline{\Omega}$, with each A_n belonging to Ω and each B_n contained in a null set N_n. Then $\bigcup N_n$ is a null set by Theorem 2.9 (b) and it contains $\bigcup B_n$. So $\bigcup(A_n \cup B_n) = \bigcup A_n \cup \bigcup B_n$ also belongs to $\overline{\Omega}$. This shows that $\overline{\Omega}$ is stable under countable unions.

For stability under complementation, let $A \cup B$ belong to $\overline{\Omega}$ with $A \in \Omega$ and B contained in a null set N. Then $(A \cup B)^c = (A \cup N)^c \cup B'$ where $B' = N \setminus (A \cup B) \subseteq N$, so it also belongs to $\overline{\Omega}$ (see Figure 2.1). We conclude that $\overline{\Omega}$ is a σ-algebra.

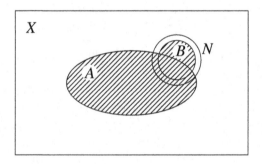

Fig. 2.1 $(A \cup B)^c = (A \cup N)^c \cup B'$

Next, we must check that $\overline{\mu}$ is well-defined. Suppose $A \cup B = A' \cup B'$ with $A, A' \in \Omega$ and $B \subseteq N$, $B' \subseteq N'$ with N and N' null sets. Then $A \setminus (N \cup N') = A' \setminus (N \cup N')$ and hence $\mu(A) = \mu(A')$. Thus $\overline{\mu}$ is well-defined. To see that it is a measure, suppose $(A_n \cup B_n)$ is a sequence of disjoint sets in $\overline{\Omega}$; then the A_n are also disjoint, so

$$\overline{\mu}\left(\bigcup(A_n \cup B_n)\right) = \mu\left(\bigcup A_n\right) = \sum \mu(A_n) = \sum \overline{\mu}(A_n \cup B_n).$$

So $\overline{\mu}$ is a measure. Finally, if $\overline{\mu}(A \cup B) = 0$ and B is contained in the null set N, then $A \cup N$ is also a null set and any subset of $A \cup B$ is contained in $A \cup N$, and therefore belongs to $\overline{\Omega}$. This shows that $\overline{\mu}$ is complete. $\square$

2.3 Premeasures

As we have already noted, the stability of σ-algebras under countable set-theoretic operations is a very strong requirement, so strong that in typical examples all subsets of interest are measurable. The cost of this feature is that it makes it difficult to construct measures. Except in very special cases, we cannot simply "write down" a measure on a σ-algebra in any really direct way. Instead, the general technique is to start with some partial description of a measure on a more manageable family of sets and then use an approximation argument to extend it to a σ-algebra. Fortunately, we can codify such strategies so that they do not have to be reinvented for each new example. We will do this now. A different method will appear in Section 4.5.

Definition 2.12. A *ring of subsets* of a set X is a family Ω_0 of subsets of X which includes $\emptyset$ and is stable under set-theoretic differences and finite unions. A *premeasure* on X is a function μ_0 from such a ring Ω_0 into $[0, \infty)$ satisfying $\mu_0(\emptyset) = 0$ and $\mu_0(\bigcup A_n) = \sum \mu_0(A_n)$ whenever (A_n) is a sequence of disjoint sets in Ω_0 whose union also belongs to Ω_0.

Note that we require premeasures to take values in $[0, \infty)$; infinite measure sets are not allowed. This is why we ask the family of sets on which a premeasure is defined to be stable under differences but not complements. Although $\emptyset$ belongs to every ring of sets, the whole set X need not belong.

The additivity condition deserves some explanation. Since Ω_0 is only assumed to be stable under finite set-theoretic operations, the infinite union $\bigcup_{n=1}^{\infty} A_n$ need not belong to Ω_0. However, if it happens that for some sequence of disjoint sets (A_n) the union $\bigcup A_n$ does belong to Ω_0, then μ_0 cannot possibly extend to a measure if it fails to be additive on this sequence. We can summarize by saying that we require a premeasure to be countably additive to the extent this makes sense.

One situation when we can be sure that the infinite union does belong to Ω_0 is when all but finitely many of the sets A_n are empty. In that case the infinite union becomes a finite union and the finite stability properties of a ring of sets apply. It follows that every premeasure is finitely additive.

Given a premeasure μ_0 defined on a ring Ω_0, we define its associated *outer measure* μ^* by

$$\mu^*(A) = \inf \left\{ \sum_{n=1}^{\infty} \mu_0(A_n) : \text{ each } A_n \in \Omega_0 \text{ and } A \subseteq \bigcup_{n=1}^{\infty} A_n \right\}$$

for any subset $A \subseteq X$, using the convention that $\inf \emptyset = \infty$. Roughly speaking, $\mu^*(A)$ is the minimal measure of a covering of A by countably many sets in Ω_0. In general μ^* is not countably additive, but we will show that there is always a σ-algebra that contains the original ring Ω_0 on which μ^* is countably additive.

Let $A \triangle B$ denote the symmetric difference of A and B, i.e., $A \triangle B = (A \setminus B) \cup (B \setminus A)$.

Lemma 2.13. *Let X be a set, let μ_0 be a premeasure on X with domain Ω_0, and let μ^* be the associated outer measure. Then*

(i) $A \subseteq B$ implies $\mu^*(A) \leq \mu^*(B)$ for all $A, B \subseteq X$;
(ii) $\mu^*(\bigcup A_n) \leq \sum \mu^*(A_n)$ for any sequence (A_n) of subsets of X;
(iii) $|\mu^*(A) - \mu^*(B)| \leq \mu^*(A \triangle B)$ for all $A, B \subseteq X$;
(iv) $\mu^*(A) = \mu_0(A)$ for all $A \in \Omega_0$.

Proof. That $A \subseteq B$ implies $\mu^*(A) \leq \mu^*(B)$ is trivial, since any cover of B will also be a cover of A and hence anything in the infimum that defines $\mu^*(B)$ will also contribute to the infimum that defines $\mu^*(A)$. Next, if (A_n) is any sequence of subsets of X and $\epsilon > 0$, then for each n we can find a cover (A_n^i) of A_n such that

$$\sum_{i=1}^{\infty} \mu_0(A_n^i) \leq \mu^*(A_n) + \frac{\epsilon}{2^n},$$

and then the family $\{A_n^i\}$ is a cover of $\bigcup A_n$ which verifies

$$\mu^* \left(\bigcup_n A_n \right) \leq \sum_{i,n} \mu_0(A_n^i) \leq \sum_n \left(\mu^*(A_n) + \frac{\epsilon}{2^n} \right) = \epsilon + \sum_n \mu^*(A_n).$$

Taking $\epsilon \to 0$ completes the proof of property (ii).

Property (iii) follows directly from properties (i) and (ii): we have $A \subseteq (A \triangle B) \cup B$, so $\mu^*(A) \leq \mu^*((A \triangle B) \cup B) \leq \mu^*(A \triangle B) + \mu^*(B)$, and similarly $\mu^*(B) \leq \mu^*(A \triangle B) + \mu^*(A)$. Thus $|\mu^*(A) - \mu^*(B)| \leq \mu^*(A \triangle B)$.

Finally, let $A \in \Omega_0$. Then taking $A_1 = A$ and $A_n = \emptyset$ for $n > 1$ shows that $\mu^*(A) \leq \mu_0(A)$. Conversely, let (A_n) be any sequence in Ω_0 whose union contains A. Define $B_1 = A \cap A_1$ and

$$B_n = A \cap \left(A_n \setminus (A_1 \cup \cdots \cup A_{n-1}) \right)$$

for $n > 1$. Then the sets B_n are disjoint and their union equals A, so $\mu_0(A) = \sum \mu_0(B_n)$ by the definition of a premeasure. Thus

$$\mu_0(A) = \sum \mu_0(B_n) \leq \sum \mu_0(A_n),$$

and we conclude that $\mu_0(A) \leq \mu^*(A)$. So $\mu^*(A) = \mu_0(A)$. $\qquad \square$

That is, μ^* is monotone and subadditive, it satisfies a triangle inequality, and it agrees with μ_0 on Ω_0.

Now we show how to get a measure from a premeasure. The idea of the following proof is that μ^* becomes a measure when it is restricted to the right σ-algebra, and the right σ-algebra consists of those sets A which are approximated by sets $A' \in \Omega_0$ in the sense that $\mu^*(A \triangle A')$ can be made arbitrarily small. This strategy requires $X \in \Omega_0$, so if $X \notin \Omega_0$ then we first work within sets in the ring and then take a supremum.

Theorem 2.14. *Any premeasure extends to a complete measure.*

Proof. Let μ_0 be a premeasure on a set X with domain Ω_0. In the first part of the proof we assume that X belongs to Ω_0. Let μ^* be the associated outer measure and let Ω be the family of subsets $A \subseteq X$ such that for every $\epsilon > 0$ there exists $A' \in \Omega_0$ with $\mu^*(A \triangle A') \le \epsilon$. We will show that Ω is a σ-algebra and the restriction of μ^* to Ω is a complete measure. It is clear that Ω contains Ω_0, and we already know that μ^* agrees with μ_0 on Ω_0.

We start by showing that Ω is stable under complements and finite unions and that μ^* is finitely additive on Ω. Stability of Ω under complements follows from the fact that $A \triangle A' = A^c \triangle (A')^c$, so that $(A')^c$ approximates A^c exactly as well as A' approximates A. For stability under finite unions let $A, B \in \Omega$ and let $\epsilon > 0$. Find $A', B' \in \Omega_0$ such that $\mu^*(A \triangle A'), \mu^*(B \triangle B') \le \epsilon$. Then $A' \cup B' \in \Omega_0$ and

$$\mu^*\big((A \cup B)\triangle(A' \cup B')\big) \le \mu^*\big((A \triangle A') \cup (B \triangle B')\big)$$
$$\le \mu^*(A \triangle A') + \mu^*(B \triangle B') \le 2\epsilon$$

since μ^* is monotone and subadditive. This shows that $A \cup B \in \Omega$. Moreover, if A and B are disjoint then $A' \cap B' \subseteq (A \triangle A') \cup (B \triangle B')$, so

$$\mu^*(A' \cap B') \le \mu^*\big((A \triangle A') \cup (B \triangle B')\big) \le 2\epsilon.$$

Since μ^* agrees with μ_0, which is additive, on Ω_0, we therefore have

$$\mu^*(A') + \mu^*(B') - \mu^*(A' \cup B') = \mu^*(A' \cap B') \le 2\epsilon.$$

Using Lemma 2.13 (iii) to bound $|\mu^*(A) - \mu^*(A')|$, $|\mu^*(B) - \mu^*(B')|$, and $|\mu^*(A \cup B) - \mu^*(A' \cup B')|$ then yields

$$\mu^*(A) + \mu^*(B) - \mu^*(A \cup B) \le 6\epsilon.$$

We conclude that μ^* is finitely additive on Ω.

We now show that Ω is a σ-algebra using Proposition 2.2. Thus, let (A_n) be a sequence of disjoint sets in Ω and let $A = \bigcup A_n$. Since μ^* is finitely additive on Ω we must have $\sum_{n=1}^{N} \mu^*(A_n) = \mu^*(\bigcup_{n=1}^{N} A_n) \le \mu^*(X)$ for all N, and hence $\sum_{n=1}^{\infty} \mu^*(A_n) \le \mu^*(X)$. So for any $\epsilon > 0$ we can find N such that $\sum_{n=N+1}^{\infty} \mu^*(A_n) \le \epsilon$. Note that this also implies $\mu^*(\bigcup_{n=N+1}^{\infty} A_n) \le \epsilon$ by countable subadditivity. Now since Ω is stable under finite unions we can find $A' \in \Omega_0$ such that $\mu^*((A_1 \cup \cdots \cup A_N) \triangle A') \le \epsilon$, and we then have $\mu^*(A \triangle A') \le 2\epsilon$, which shows that $A \in \Omega$. So Ω is stable under countable disjoint unions, and therefore by Proposition 2.2 it is a σ-algebra. Moreover, since μ^* is finitely additive we also have

$$\left| \mu^*(A) - \sum_{n=1}^{\infty} \mu^*(A_n) \right| \le \left| \mu^*(A) - \mu^* \left(\bigcup_{n=1}^{N} A_n \right) \right| + \sum_{n=N+1}^{\infty} \mu^*(A_n)$$

$$= \mu^* \left(\bigcup_{n=N+1}^{\infty} A_n \right) + \sum_{n=N+1}^{\infty} \mu^*(A_n) \le 2\epsilon.$$

We conclude that μ^* is countably additive on Ω.

To establish completeness, just observe that $\mu^*(A) = 0$ implies that $A \in \Omega$ by taking $A' = \emptyset$. Since μ^* is monotone, this shows that any subset of a null set belongs to Ω.

This completes the proof in the case that $X \in \Omega_0$. In the general case, for each $A \in \Omega_0$ let Ω_0^A be the family of sets in Ω_0 which are contained in A and let μ_0^A be the restriction of μ_0 to Ω_0^A. Then the argument given above extends μ_0^A to a measure μ^A defined on a σ-algebra Ω^A of subsets of A which contains Ω_0^A. Now let Ω be the family of subsets B of X with the property that $B \cap A \in \Omega^A$ for every $A \in \Omega_0$. It is more or less immediate that Ω is a σ-algebra of subsets of X which contains Ω_0. For each $B \in \Omega$ define

$$\mu(B) = \sup_{A \in \Omega_0} \mu^A(B \cap A).$$

The key observation here is that if B is contained in both A_1 and A_2 then $\mu^{A_1}(B) = \mu^{A_2}(B)$. To see this, observe that any sequence of sets in $\Omega_0^{A_1}$ which covers B can be intersected with A_2 to become a sequence of sets in $\Omega_0^{A_2}$ which covers B and has smaller total measure, and vice versa. This observation entails that μ agrees with μ_0 on Ω_0. Also, given any sequence of disjoint sets B_n in Ω, we have

$$\mu^A \left(\bigcup (B_n \cap A) \right) = \sum \mu^A(B_n \cap A)$$

for all $A \in \Omega_0$, and taking the supremum of both sides over A yields $\mu(\bigcup B_n) = \sum \mu(B_n)$, i.e., μ is countably additive. So μ is a measure. Completeness of μ follows easily from completeness of each μ^A. $\square$

2.4 Lebesgue measure

We now have a general method for generating measures. It can be used in a variety of situations; for our first application we will construct Lebesgue (volume) measure on $\mathbf{R}^n$.

Let a *half-open box* in $\mathbf{R}^n$ be a set of the form $(a_1, b_1] \times \cdots \times (a_n, b_n]$ where $a_i < b_i$ for each i, and define the *volume* of this box to be $(b_1 - a_1) \cdot \ldots \cdot (b_n - a_n)$. Half-open boxes have the nice property that the set-theoretic difference of any two half-open boxes can be expressed as a finite union of disjoint half-open boxes. If A is a finite union of disjoint half-open boxes, let $m_0(A)$ be the sum of the volumes of the component boxes.

We sketch a proof that $m_0(A)$ is well-defined. The issue is that the decomposition of a set into disjoint half-open boxes can generally be accomplished in more than one way. But any two such decompositions have a common refinement (just intersect every box in the first decomposition with every box in the second decomposition; the intersection of any two half-open boxes is a half-open box), so it is enough to show that the total volume of any finite union of disjoint half-open boxes equals the total volume of any refinement of it. By considering each box in the former union separately, we need only show that the volume of a single box equals the total volume of any refinement of it. Even this might not be obvious, because a box can be broken up into smaller boxes in complicated ways. But in the special case where each side of the box is partitioned into subintervals and we form the smaller boxes by taking, in all possible ways, products of one subinterval from each axis, the consistency of the volume calculation is a routine (though notationally tedious) exercise. This is enough because

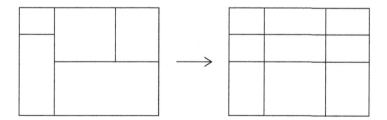

Fig. 2.2   Refining a partition of a box

however a box is decomposed into finitely many smaller boxes, it will be possible to refine the partition further so that it has this special form (see Figure 2.2).

The finite unions of disjoint half-open boxes constitute a ring of subsets of $\mathbf{R}^n$. This is a consequence of the following lemma.

Lemma 2.15. *Let S be a family of subsets of a set X and suppose that the difference of any two sets in S is a finite union of disjoint sets in S. Then the collection of all finite unions of disjoint sets in S is a ring of subsets of X.*

Proof. Let Ω_0 be the collection of all finite unions of disjoint sets in S. This includes the union of no sets, which is the empty set, so we have $\emptyset \in \Omega_0$. To see that Ω_0 is stable under differences, let $A_1 \cup \cdots \cup A_m$ and $B_1 \cup \cdots \cup B_n$ be finite unions of disjoint sets in S; we must show that their difference can be expressed as a finite union of disjoint sets in S. This holds for $n = 1$ because

$$(A_1 \cup \cdots \cup A_m) \setminus B_1 = (A_1 \setminus B_1) \cup \cdots \cup (A_m \setminus B_1);$$

the differences on the right are all disjoint since the A_i are disjoint, and by hypothesis each of them can be expressed as a finite union of disjoint sets in S, so we achieve the desired expression. For $n > 1$ we use the identity

$$A \setminus (B_1 \cup \cdots \cup B_n) = ((A \setminus B_1) \setminus \cdots) \setminus B_n$$

and invoke the $n = 1$ case repeatedly.

We now know that Ω_0 is stable under differences, and it is trivially stable under finite disjoint unions; these two facts together imply that it is stable under finite unions since $A \cup B = A \cup (B \setminus A)$. So we conclude that Ω_0 is a ring of sets. $\qquad\square$

Thus, the domain of the volume function m_0 is a ring. To be able to apply Theorem 2.14 we just have to check that m_0 is a premeasure. In the following proof we refer to open and closed boxes, which are defined analogously to half-open boxes. We also define the volume of an open or closed box to be the product of the lengths of the sides, just as for half-open boxes.

Proposition 2.16. *The volume function m_0 is a premeasure on the ring of finite unions of disjoint half-open boxes.*

Proof. Let Ω_0 be the ring of finite unions of disjoint half-open boxes. Suppose (A_n) is a sequence of disjoint sets in Ω_0 whose union A also belongs to Ω_0; we must show that $m_0(A) = \sum m_0(A_n)$. It is immediate from the definition of m_0 that m_0 is finitely additive, and as in Theorem 2.9 (a) this implies that m_0 is monotone. So for any $N \in \mathbf{N}$ we have

$$\sum_{n=1}^{N} m_0(A_n) = m_0 \left(\bigcup_{n=1}^{N} A_n \right) \leq m_0(A),$$

and taking $N \to \infty$ yields $\sum m_0(A_n) \leq m_0(A)$.

For the reverse inequality, observe that any half-open box contains a closed box of arbitrarily near volume and is contained in an open box of arbitrarily near volume. Since $A \in \Omega_0$, it follows that given $\epsilon > 0$ we can find a finite union B of disjoint closed boxes which is contained in A and the sum of whose volumes, V, equals $m_0(A) - \epsilon$. Similarly, for each n we can find a finite union B_n of open boxes which contains A_n and the sum of whose volumes, V_n, equals $m_0(A_n) + \frac{\epsilon}{2^n}$. Then the B_n cover B and by compactness we can find a finite subcover. So B is contained in $\bigcup_{n=1}^{N} B_n$ for some N. Now if we replace all of the closed boxes that make up B and open boxes that make up the B_n with half-open boxes, then we get sets B' and B'_n in Ω_0 which satisfy $B' \subset \bigcup_{n=1}^{N} B'_n$, $m_0(B') = V = m_0(A) - \epsilon$, and

$$m_0 \left(\bigcup_{n=1}^{N} B'_n \right) \leq \sum_{n=1}^{\infty} V_n = \sum_{n=1}^{\infty} m_0(A_n) + \epsilon,$$

so that $m_0(A) - \epsilon \leq \sum m_0(A_n) + \epsilon$. Taking $\epsilon \to 0$ yields the desired inequality. $\qquad\square$

Definition 2.17. The measure m on $\mathbf{R}^n$ generated via Theorem 2.14 from the premeasure m_0 is *Lebesgue measure*. Its domain is the *Lebesgue σ-algebra* and a set in this σ-algebra is said to be *Lebesgue measurable* (or just *Lebesgue*).

The Lebesgue σ-algebra on $\mathbf{R}^n$ is easily seen to contain every open box and hence every open set. We will get a better idea of what arbitrary Lebesgue sets look like in Theorem 2.19 (ii).

Example 2.18. Let $\mathbf{a} = (a_1, \ldots, a_n) \in \mathbf{R}^n$. Then $\{\mathbf{a}\} \subset (a_1 - \epsilon, a_1] \times \cdots \times (a_n - \epsilon, a_n]$ for any $\epsilon > 0$. By monotonicity, we have $m(\{\mathbf{a}\}) \leq \epsilon^n$ for all ϵ, and hence $m(\{\mathbf{a}\}) = 0$. By countable additivity, the Lebesgue measure of any countable subset of $\mathbf{R}^n$ is zero.

Incidentally, this yields a new proof that $[0, 1]$ is uncountable, since its Lebesgue measure is 1.

A G_δ set is a countable intersection of open sets.

Theorem 2.19. *Let $A \in \mathbf{R}^n$ be Lebesgue measurable. Then*

(i) $m(A) = \inf\{m(U) : A \subseteq U$ *and* U *is open*$\} = \sup\{m(K) : K \subseteq A$ *and*
   K *is compact*$\}$;

(ii) *we can write* $A = A' \setminus N$ *where* A' *is* G_δ *and* N *is a null set;*

(iii) *if* $m(A) < \infty$ *then for any* $\epsilon > 0$ *there is a finite union of open boxes*
   B *such that* $m(A \triangle B) < \epsilon$.

Proof. (i) If $A \subseteq U$ then $m(A) \leq m(U)$. Conversely, if A is bounded then it is contained in a half-open box, and given $\epsilon > 0$, by the construction of m we can find a sequence of half-open boxes B_n such that $A \subseteq \bigcup B_n$ and $m(A) \geq \sum m(B_n) - \epsilon$. Then for each n we can find an open box U_n which contains B_n and whose measure is $\frac{\epsilon}{2^n}$ greater, and we get that $A \subseteq \bigcup U_n$ and $m(A) \geq \sum m(U_n) - 2\epsilon$, which is enough to prove the first equality. If A is unbounded it can be decomposed into a countable union of disjoint bounded sets, and by the preceding we can cover these sets with a sequence of open sets in such a way that the total discrepancy in measure is at most ϵ. This suffices to prove the first equality in general.

For the second equality, observe that $m(A) = \lim m(A \cap \mathrm{ball}_N(0))$, so we can assume A is bounded. Let $\epsilon > 0$, let U be an open set containing $\overline{A} \setminus A$ such that $m(U) \leq m(\overline{A} \setminus A) + \epsilon$, and let $K = \overline{A} \setminus U$. Then K is compact, $K \subseteq A$, and

$$m(K) \geq m(\overline{A}) - m(U) \geq m(\overline{A}) - m(\overline{A} \setminus A) - \epsilon = m(A) - \epsilon.$$

This suffices to prove the second equality.

(ii) First suppose $m(A) < \infty$. For each $n \in \mathbf{N}$ find an open set U_n such that $A \subseteq U_n$ and $m(A) \geq m(U_n) - \frac{1}{n}$. Then $A' = \bigcap U_n$ is a G_δ set which contains A, and $m(A') \leq m(U_n) \leq m(A) + \frac{1}{n}$ for all n, so $m(A') = m(A)$. Thus $N = A' \setminus A$ is a null set, and we have $A = A' \setminus N$.

If $m(A) = \infty$, for each $k, n \in \mathbf{N}$ find an open set U_n^k such that $A \cap (\mathrm{ball}_{k+1}(0) \setminus \mathrm{ball}_k(0)) \subseteq U_n^k$ and

$$m\big(A \cap (\mathrm{ball}_{k+1}(0) \setminus \mathrm{ball}_k(0))\big) \geq m(U_n^k) - \frac{1}{2^k n}.$$

Let $A' = \bigcap_n \bigcup_k U_n^k$ and proceed as before.

(iii) Assume $m(A) < \infty$ and as in part (a) find a sequence (U_n) of open boxes such that $A \subseteq \bigcup U_n$ and $m(A) \geq \sum m(U_n) - 2\epsilon$. Since $\bigcup U_n$ is a countable union of open boxes, for sufficiently large N the set $B = \bigcup_{n=1}^{N} U_n$

is a finite union of open boxes which satisfies $m(B) \geq m(\bigcup U_n) - \epsilon$. We then have

$$m(A \triangle B) = m(A \setminus B) + m(B \setminus A) \leq \epsilon + 2\epsilon = 3\epsilon,$$

which is enough. □

The property described in Theorem 2.19 (i) is known as "regularity".

Another nice property of Lebesgue measure is that it behaves well under affine transformations of $\mathbf{R}^n$. Our proof of this result uses two facts from linear algebra: first, that the determinant is multiplicative (i.e., we have $\det(ST) = \det(S)\det(T)$ for any square matrices S and T), and second, that any invertible square matrix can be converted into the identity matrix by a string of operations of the following types: multiply a row by a nonzero scalar; subtract one row from another; interchange two rows. Another way to say this is that any invertible square matrix can be written as a product of matrices corresponding to maps of the three types

$$(x_1, \ldots, x_j, \ldots, x_n) \mapsto (x_1, \ldots, ax_j, \ldots, x_n)$$
$$(x_1, \ldots, x_j, \ldots, x_n) \mapsto (x_1, \ldots, x_j - x_k, \ldots, x_n)$$
$$(x_1, \ldots, x_j, \ldots, x_k, \ldots, x_n) \mapsto (x_1, \ldots, x_k, \ldots, x_j, \ldots, x_n).$$

Theorem 2.20. *Lebesgue measure on* $\mathbf{R}^n$ *is translation invariant. If* $T : \mathbf{R}^n \to \mathbf{R}^n$ *is an invertible linear map and* $A \subseteq \mathbf{R}^n$ *is Lebesgue measurable, then* $T(A)$ *is Lebesgue measurable and* $m(T(A)) = |\det(T)|m(A)$. *Lebesgue measure is invariant under the action of the orthogonal group* $O(n)$.

Proof. For translation invariance, let $\mathbf{x} \in \mathbf{R}^n$ and define $T_{\mathbf{x}} : \mathbf{R}^n \to \mathbf{R}^n$ by $T_{\mathbf{x}}\mathbf{y} = \mathbf{y} + \mathbf{x}$. Then $T_{\mathbf{x}}$ takes the ring generated by half-open boxes to itself, and $T_{\mathbf{x}}$ preserves the premeasure on this ring. It follows that $T_{\mathbf{x}}$ preserves the measure constructed from this premeasure via Theorem 2.14. Thus Lebesgue measure is invariant under $T_{\mathbf{x}}$.

Since the determinant is multiplicative, by the comment preceding the theorem it suffices to prove that $T(A)$ is measurable and that $m(T(A)) = |\det(T)|m(A)$ only for maps T of the three types listed there. It is easy to check this for maps of the first and third types by first checking it on the ring generated by the half-open boxes. For a map T of the second type, first observe that the assertion holds for any half-open box because the image of a half-open box under T is a skew half-open box which can be decomposed into finitely many pieces and reassembled to form the original box. Since any open set is a union of countably many disjoint half-open boxes (first write the set as a union of countably many overlapping half-open boxes, then disjointify), the assertion then holds for all open sets, and by intersecting a descending sequence of open sets we get it for any G_δ set. Finally, we observe that the assertion holds for any subset of a null G_δ set, and since every null set is countained in a null G_δ and every Lebesgue set is a G_δ set minus a null set (Theorem 2.19 (ii)), the full assertion follows.

$T \in O(n)$ if $\det(T) = \pm 1$, so the last assertion follows immediately. $\square$

We can use translation invariance to prove that there exist sets which are not Lebesgue measurable.

Example 2.21. (Vitali) Define an equivalence relation on $[0, 1)$ by setting $s \sim t$ if $t - s \in \mathbf{Q}$, and let A be a subset of $[0, 1)$ which contains exactly one element from each equivalence class. Consider the set

$$B = \bigcup_{r \in \mathbf{Q} \cap [0,1)} A + r.$$

Observe that the sets $A+r$ are disjoint: if $s \in (A+r_1) \cap (A+r_2)$ then $s-r_1$ and $s-r_2$ both belong to A, but their difference is rational, so we must have $r_1 = r_2$. Also, observe that $[0, 1)$ is the disjoint union of $B_1 = B \cap [0, 1)$ and $B_2 = (B \cap [1, 2)) - 1$: these sets are disjoint by an argument similar to the one which showed the sets $A + r$ are disjoint, and every element s of $[0, 1)$ is equivalent to some $t \in A$, yielding $s \in A + (s - t)$ if $s \geq t$ and $s \in A + (1 + s - t) - 1$ if $s < t$.

Suppose A were Lebesgue measurable. Then by translation invariance,

$$m(B) = \sum_{r \in \mathbf{Q} \cap [0,1)} m(A + r) = \begin{cases} 0 & \text{if } m(A) = 0 \\ \infty & \text{if } m(A) > 0. \end{cases}$$

But $m(B) = m(B_1) + m(B_2) = m([0, 1)) = 1$, a contradiction. We conclude that A cannot be measurable.

An even worse phenomenon occurs in $\mathbf{R}^3$: according to the Banach-Tarski paradox, there is a set A in $\mathbf{R}^3$ such that $\text{ball}_1(0)$ is a union of two sets, both congruent to A, and it is also a union of three sets, each congruent to A.

2.5   Lebesgue integration

We can use a measure on a set to integrate measurable scalar-valued functions. The idea is that in order to integrate a function we need to know how much weight to assign to regions where the function takes different values, and a measure provides this information. Conversely, if we can integrate measurable functions then integrating the characteristic function of a measurable subset recovers the measure of that set. So measure and integration are effectively equivalent.

We define the integrals of positive functions first. The positive case is easier because convergence is less of an issue in the same way that this is true of infinite series (any sum of positive terms either converges to a finite value or diverges to infinity, and in either case rearranging the terms has no effect). Once we can integrate positive functions, we will define the integrals of real and complex functions by expressing them as linear combinations of positive functions.

When we treat the positive case it is convenient to allow functions to take the value $+\infty$. We can define measurability of functions into $[0, \infty]$ via Definition 2.3, giving $[0, \infty]$ its natural topology according to which it is homeomorphic to $[0, 1]$. Then the analog of Lemma 2.4 states that $f : X \to [0, \infty]$ is measurable if and only if $f^{-1}((a, \infty])$ is a measurable subset of X for all $a \in [0, \infty)$. Theorem 2.6 goes through as well, with the assumption of pointwise boundedness in part (d) no longer being necessary and with sums and products defined by setting $a + \infty = \infty + a = \infty$ for all $a \in [0, \infty]$, $a \cdot \infty = \infty \cdot a = \infty$ for all $a \in (0, \infty]$, and $0 \cdot \infty = \infty \cdot 0 = 0$. (Showing that products of measurable functions are measurable requires a slightly different proof because the product function from $[0, \infty]^2$ to $[0, \infty]$ is not continuous, but this is not so important for us because we will not need to take products of functions into $[0, \infty]$.)

We denote the characteristic function of a set A by 1_A. Thus $1_A(x) = 1$ if $x \in A$ and 0 if $x \notin A$.

Definition 2.22. Let (X, μ) be a measure space. A *simple function* on X is a measurable function $h : X \to \mathbf{F}$ with finite range. (Note that we do not allow simple functions to take the value ∞.) Every such function has a *standard form* $h = \sum_{i=1}^{k} a_i \cdot 1_{A_i}$ with distinct coefficients a_i such that the A_i are disjoint and $\bigcup A_i = X$. The standard form is unique up to the order in which the terms are summed.

Let $f : X \to [0, \infty]$ be measurable. We define the *Lebesgue integral* of f, denoted $\int f$ or $\int f \, d\mu$, to be the (possibly infinite) value

$$\int f \, d\mu = \sup \left\{ \sum a_i \mu(A_i) \right\}$$

where the supremum is taken over all simple functions h satisfying $0 \leq h \leq f$ and whose standard form is $h = \sum a_i \cdot 1_{A_i}$.

The idea of this definition is that the integral of a simple function $\sum a_i \cdot 1_{A_i}$ should be $\sum a_i \mu(A_i)$, and we can leverage that formula into a definition of the integral of any positive measurable function by letting $\int f$ be the supremum of $\int h$ over all positive simple functions $h \leq f$. We use the notion of standard form to avoid having to deal with possible nonuniqueness issues.

When evaluating the product $a_i \cdot 1_{A_i}$ we use the convention $0 \cdot \infty = \infty \cdot 0 = 0$. This is the appropriate convention for measure theory because an infinite vertical ray in the plane with zero width has zero area. We can see this by covering the ray with a sequence of rectangles of height 1 and width $\frac{\epsilon}{2^n}$, for arbitrary $\epsilon > 0$. The total area in this sequence is $\sum \frac{\epsilon}{2^n} = \epsilon$, and it covers the entire ray; since ϵ was arbitrary, by monotonicity the ray must have measure zero. (It then follows from Theorem 2.20 that every infinite ray in $\mathbf{R}^2$ has measure zero.)

For $A \subseteq X$ and $f : X \to [0, \infty]$ measurable, we define $\int_A f \, d\mu$ to be the integral of $f \cdot 1_A$.

We start by noting that Lebesgue integration subsumes ordinary summation.

Example 2.23. Equip $\mathbf{N}$ with counting measure. Then every function $f : \mathbf{N} \to [0, \infty]$ is measurable, and we have $\int f \, d\mu_{\text{count}} = \sum_{n=1}^{\infty} f(n)$.

Next we verify some basic properties of integrals of simple functions.

Lemma 2.24. *Let* (X, μ) *be a measure space and let* $h, k \geq 0$ *be simple functions with standard forms* $h = \sum a_i \cdot 1_{A_i}$ *and* $k = \sum b_j \cdot 1_{B_j}$.

(a) If $h \leq k$ *then* $\sum a_i \mu(A_i) \leq \sum b_j \mu(B_j)$.
(b) We have $\int h = \sum a_i \mu(A_i)$.
(c) We have $\int (ah + bk) = a \int h + b \int k$ *for all* $a, b \geq 0$.
(d) The map $A \mapsto \int_A h$ *is a measure on* X.

Proof.   (a) Suppose $h \leq k$. Then by finite additivity of μ we have
$$\sum_i a_i \mu(A_i) = \sum_{i,j} a_i \mu(A_i \cap B_j)$$
and
$$\sum_j b_j \mu(B_j) = \sum_{i,j} b_j \mu(A_i \cap B_j).$$
On the set $A_i \cap B_j$ the functions h and k are constantly equal to a_i and b_j respectively, so $h \leq k$ implies that if $A_i \cap B_j \neq \emptyset$ then $a_i \leq b_j$. Thus
$$\sum_{i,j} a_i \mu(A_i \cap B_j) \leq \sum_{i,j} b_j \mu(A_i \cap B_j),$$

implying that $\sum_i a_i \mu(A_i) \le \sum_j b_j \mu(B_j)$.

(b) This follows directly from part (a) and the definition of the integral.

(c) We have $ah + bk = \sum_{i,j}(a \cdot a_i + b \cdot b_j) 1_{A_i \cap B_j}$. This might not be the standard form of $ah + bk$, but since the sets $A_i \cap B_j$ are disjoint it is at worst a refinement of the standard form. So using part (b) and several applications of finite additivity of μ we get

$$\int (ah + bk) = \sum_{i,j}(a \cdot a_i + b \cdot b_i)\mu(A_i \cap B_j) = a \int h + b \int k.$$

(d) It is clear that $\int_\emptyset h = 0$. If (C_n) is a sequence of disjoint measurable sets, let $C = \bigcup C_n$; then

$$\int_C h = \sum_i a_i \mu(C \cap A_i) = \sum_{i,n} a_i \mu(C_n \cap A_i) = \sum_n \int_{C_n} h,$$

which verifies countable additivity. $\qquad\square$

It follows from part (c) that the integral of a simple function $\sum a_i \cdot 1_{A_i}$ equals $\sum a_i \mu(A_i)$ regardless of whether this is the standard form.

The next result, the monotone convergence theorem (MCT), is the fundamental theorem about convergence of integrals of positive functions. Observe that $f \le g$ implies $\int f \le \int g$; this is obvious from the definition of the integral.

Theorem 2.25. *(Monotone convergence theorem) Let (X, μ) be a measure space and let (f_n) be a pointwise increasing sequence of measurable functions from X into $[0, \infty]$. Then $\int \lim f_n = \lim \int f_n$.*

Proof. Let $f = \lim f_n$, taking the limit pointwise, and note that f is measurable by the $[0, \infty]$ version of Theorem 2.6 (e). Then $f_n \le f$ for all n, so $\int f_n \le \int f$ for all n and hence

$$\lim \int f_n = \sup \int f_n \le \int f.$$

Conversely, let $\epsilon > 0$ and let h be a simple function such that $0 \le h \le f$. It will suffice to show that $\lim \int f_n \ge (1 - \epsilon) \int h$; then taking $\epsilon \to 0$ and the supremum over h will yield $\lim \int f_n \ge \int f$.

For each n define

$$A_n = \{x : f_n(x) \ge (1 - \epsilon)h(x)\}.$$

Then (A_n) is an increasing sequence of measurable sets and $\bigcup A_n = X$. We have

$$\int f_n \ge \int_{A_n} f_n \ge (1 - \epsilon) \int_{A_n} h,$$

and taking the limit as $n \to \infty$ and using Lemma 2.24 (d) we obtain

$$\lim \int f_n \geq (1 - \epsilon) \int h,$$

as desired.   □

If the sequence (f_n) is not increasing, we can still say something.

Corollary 2.26. *(Fatou's lemma) Let (X, μ) be a measure space and let (f_n) be any sequence of measurable functions from X into $[0, \infty]$. Then $\int (\liminf f_n) \leq \liminf \int f_n$.*

Proof.   For each n let $g_n = \inf_{k \geq n} f_k$, so that the sequence (g_n) is increasing and $\liminf f_n = \lim g_n$. Thus, the monotone convergence theorem yields

$$\int (\liminf f_n) = \int \lim g_n = \lim \int g_n.$$

For each n we have $g_n \leq f_n$; this implies $\lim \int g_n \leq \liminf \int f_n$, which together with the displayed equation yields the desired inequality.   □

In particular, if the sequence (f_n) converges pointwise to f but it is not increasing, then we have $\int f \leq \liminf \int f_n$. The asymmetry in this statement (we do not claim $\int f \geq \limsup \int f_n$) arises from the assumption that all the functions are positive. Indeed, the inequality in Fatou's lemma can be strict. For example, if $f_n = 1_{[n,n+1]}$ on $\mathbf{R}$ then $f_n \to 0$ pointwise but $\liminf \int f_n = 1$. Mass can "escape" in the limit.

We can use the monotone convergence theorem to quickly deduce the basic properties of integrals of positive functions. The basic idea is to reduce to the simple case using MCT and the fact that any positive measurable function f is the pointwise limit of an increasing sequence (h_n) of simple functions. For instance, we could define

$$h_n(x) = \begin{cases} \frac{k}{2^n} & \text{if } \frac{k}{2^n} \leq f(x) < \frac{k+1}{2^n} \text{ for some } k < 4^n \\ 2^n & \text{if } f(x) \geq 2^n. \end{cases}$$

Here we use the denominator 2^n so that h_{n+1} matches up with h_n: if h_n takes the value $\frac{k}{2^n}$ at some point then h_{n+1} takes either the value $\frac{2k}{2^{n+1}} = \frac{k}{2^n}$ or the value $\frac{2k+1}{2^{n+1}}$ at that point. This ensures that $h_n \leq h_{n+1}$. Since f could be unbounded, in order for the range of h_n to be finite we have to cut it off at some point, but the cutoff point should go to infinity so that the pointwise limit of h_n equals f everywhere. In the above formula we cut off at 2^n.

Theorem 2.27. *Let (X, μ) be a measure space and let $f, g, f_n : X \to [0, \infty]$ be measurable functions.*

(a) If $f \le g$ then $\int f \le \int g$.
(b) We have $\int (af + bg) = a \int f + b \int g$ for all $a, b \ge 0$.
(c) We have $\int \sum f_n = \sum \int f_n$.
(d) The map $A \mapsto \int_A f$ is a measure on X.

Proof. (a) Trivial. (We already used this fact.)

(b) Let h_n and k_n be increasing sequences of simple functions which converge pointwise to f and g, respectively. Then $ah_n + bk_n$ increases pointwise to $af + bg$, so MCT and Lemma 2.24 (c) imply

$$\int (af + bg) = \lim \int (ah_n + bk_n) = \lim \left(a \int h_n + b \int k_n \right) = a \int f + b \int g.$$

(c) Using part (b) and MCT, we have

$$\sum_{n=1}^{\infty} \int f_n = \lim_{n \to \infty} \sum_{k=1}^{n} \int f_k = \lim_{n \to \infty} \int \sum_{k=1}^{n} f_k = \int \sum_{n=1}^{\infty} f_n.$$

(d) $\int_\emptyset f = 0$ is clear. For countable additivity let (A_n) be a sequence of disjoint measurable sets and for each n let $f_n = f \cdot 1_{A_n}$. Then by part (c)

$$\sum \int_{A_n} f = \sum \int f_n = \int \sum f_n = \int_A f$$

where $A = \bigcup A_n$. This verifies countable additivity. $\square$

One intuition for part (d) is to think of f as describing a mass density. Then $\int_A f$ is the amount of mass contained in A.

We now turn to the problem of integrating real- and complex-valued functions. This is just a matter of decomposing such functions into linear combinations of positive functions.

Write f^+ and f^- for the positive and negative parts of a real-valued function f. Thus $f^+ = \max(f, 0)$ and $f^- = \max(-f, 0)$. Since $|f| = f^+ + f^-$, we can say that $\int |f|$ is finite if and only if $\int f^+$ and $\int f^-$ are both finite. Also, for complex-valued f the inequalities

$$|\Re f|, |\Im f| \le |f| \le \sqrt{2} \max(|\Re f|, |\Im f|)$$

entail that $\int |f|$ is finite if and only if both $\int |\Re f|$ and $\int |\Im f|$ are finite. Thus the following definition makes sense.

Definition 2.28. Let (X, μ) be a measure space. A measurable function $f : X \to \mathbf{F}$ is *integrable* if $\int |f| < \infty$. If f is integrable, we define the *Lebesgue integral* of f, denoted $\int f$ or $\int f \, d\mu$, to be $\int f^+ - \int f^-$ in the real case and $\int \Re f + i \int \Im f$ in the complex case.

The condition $\int |f| < \infty$ should be familiar in the case of counting measure:

Example 2.29. A function $f : \mathbf{N} \to \mathbf{F}$ is integrable in the above sense with respect to counting measure if and only if it is absolutely summable, in which case we have $\int f \, d\mu_{\text{count}} = \sum_{n=1}^{\infty} f(n)$.

The basic properties of integrals of scalar-valued functions follow fairly straightforwardly from the corresponding properties for positive functions.

Theorem 2.30. *Let (X, μ) be a measure space and let $f, g : X \to \mathbf{F}$ be integrable. Then*

(i) $\int (f + g) = \int f + \int g;$
(ii) $\int af = a \int f$ *for all $a \in \mathbf{F}$;*
(iii) $|\int f| \leq \int |f|.$

Proof. (i) Integrability of $h = f + g$ follows from the inequality $|f + g| \leq |f| + |g|$ and Theorem 2.27 (b). In the real case we have
$$h^+ - h^- = h = f + g = f^+ - f^- + g^+ - g^-,$$
so $h^+ + f^- + g^- = h^- + f^+ + g^+$. Additivity of integrals of positive functions (Theorem 2.27 (b)) then implies
$$\int h^+ + \int f^- + \int g^- = \int h^- + \int f^+ + \int g^+,$$
and undoing the rearrangement yields $\int h = \int f + \int g$. In the complex case apply the preceding to real and imaginary parts separately.

(ii) The equality $\int af = a \int f$ holds for $a \geq 0$ by Theorem 2.27 (b). For $a = -1$, in the real case we have
$$\int -f = \int ((-f)^+ - (-f)^-) = \int (f^- - f^+) = -\int f,$$
and in the complex case we get the same result by considering real and imaginary parts separately. For $a = i$ we have
$$\int if = \int \Re(if) + i \int \Im(if) = -\int \Im f + i \int \Re f = i \int f.$$
Combining these results yields the full equality $\int af = a \int f$ for arbitrary a.

(iii) In the real case this follows easily from the fact that $|f| = f^+ + f^-$; in the complex case we need to use a trick. Suppose $\int f \neq 0$ and let $a = \frac{|\int f|}{\int f}$. Then $\int af = |\int f|$ is real and $|a| = 1$ so
$$\left| \int f \right| = \int af = \int \Re af \leq \int |f|,$$
as desired.	$\square$

The fundamental result about convergence of integrals of scalar-valued functions is the dominated convergence theorem (DCT). Unlike the MCT, we do not require the approximating sequence to be increasing; instead we demand that the entire sequence lie under an integrable umbrella.

Theorem 2.31. *(Dominated convergence theorem) Let (X, μ) be a measure space and let (f_n) be a sequence of integrable functions which converges pointwise to a function f. Suppose there is an integrable function $g \geq 0$ such that $|f_n| \leq g$ for all n. Then f is integrable and $\int f_n \to \int f$.*

Proof. By considering real and imaginary parts separately, we can reduce to the real case. Since $|f(x)| = \lim |f_n(x)| \leq g(x)$ for all x, f is integrable. Now we verify $\int f_n \to \int f$. We have $g \pm f_n \geq 0$, so by Fatou's lemma

$$\int g + \int f \leq \liminf \int (g + f_n) = \int g + \liminf \int f_n$$

and

$$\int g - \int f \leq \liminf \int (g - f_n) = \int g - \limsup \int f_n,$$

and hence $\liminf \int f_n \geq \int f \geq \limsup \int f_n$. Thus $\int f = \lim \int f_n$. $\square$

2.6   Product measures

Let (X, μ) and (Y, ν) be measure spaces. In this section we will construct a product measure on $X \times Y$ and relate integration on the product space to integration on X and Y.

If $A \subseteq X$ and $B \subseteq Y$ are measurable subsets with $\mu(A), \nu(B) < \infty$ then we call $A \times B \subseteq X \times Y$ a *finite measurable rectangle*. Observe that the difference of any two finite measurable rectangles can be expressed as a union of three disjoint finite measurable rectangles. So it follows from Lemma 2.15 that the family of finite unions of disjoint finite measurable rectangles constitutes a ring of sets. Denote this ring by Ω_0.

Define $(\mu \times \nu)(A \times B) = \mu(A)\nu(B)$ for any finite measurable rectangle $A \times B$ and use finite additivity to extend $\mu \times \nu$ to Ω_0. The proof that $\mu \times \nu$ is well-defined on Ω_0 is a straightforward generalization of the proof that m_0 is well-defined indicated at the beginning of Section 2.4. Now we must show that $\mu \times \nu$ is a premeasure.

Lemma 2.32. *Let (X, μ) and (Y, ν) be measure spaces. Then $\mu \times \nu$ is a premeasure on Ω_0.*

Proof. Suppose (C_n) is a sequence of disjoint sets in Ω_0 whose union C also belongs to Ω_0; we must show that $(\mu \times \nu)(C) = \sum (\mu \times \nu)(C_n)$. Since $C \in \Omega_0$ it is a finite union of disjoint rectangles, but by intersecting each of these rectangles with the sequence (C_n) we can reduce to the case where $C = A \times B$ is a single rectangle. Then by breaking up each C_n into its component rectangles, we can reduce to the case where each C_n is a single rectangle, say $C_n = A_n \times B_n$. Thus, suppose $A \times B = \bigcup_{n=1}^{\infty}(A_n \times B_n)$ where the $A_n \times B_n$ are disjoint; we must show that $\mu(A)\nu(B) = \sum_{n=1}^{\infty} \mu(A_n)\nu(B_n)$.

For each n let $f_n : X \to [0, \infty)$ be the function which takes the value $\nu(B_n)$ on the set A_n and the value 0 off the set A_n. That is, $f_n = \nu(B_n) \cdot 1_{A_n}$. Then $\int f_n \, d\mu = \mu(A_n)\nu(B_n)$. For any point $x \in A$ we have $\sum_{n=1}^{\infty} f_n(x) = \nu(B)$ since $\{B_n : x \in A_n\}$ is a sequence of disjoint sets whose union equals B. Also, for $x \notin A$ we have $f_n(x) = 0$ for all n. Thus $\sum f_n = \nu(B) \cdot 1_A$ (pointwise), so by Theorem 2.27 (c)

$$\sum_{n=1}^{\infty} \mu(A_n)\nu(B_n) = \sum_{n=1}^{\infty} \int f_n \, d\mu = \int \nu(B) \cdot 1_A \, d\mu = \mu(A)\nu(B),$$

as desired. $\square$

It is an interesting feature of this proof that we use integration theory, specifically the monotone convergence theorem as expressed in Theorem 2.27 (c), to prove a fact about pure measure theory. On the other hand, maybe this is not so surprising because integrating positive functions on X can be interpreted as evaluating volume in the product space $X \times [0, \infty]$.

It now follows from Theorem 2.14 that we can extend $\mu \times \nu$ to a complete measure on $X \times Y$. We use the same notation $\mu \times \nu$ for this extension.

Theorem 2.33. *Let (X, μ) and (Y, ν) be measure spaces. Then there is a complete measure $\mu \times \nu$ on $X \times Y$ whose domain contains all finite measurable rectangles and which satisfies $(\mu \times \nu)(A \times B) = \mu(A)\nu(B)$ for any finite measurable rectangle $A \times B$.*

The measure $\mu \times \nu$ is called a *product measure*. It is routine to extend this construction to any finite number of factors and to see that the order in which the factors are taken is immaterial. If the number of factors is infinite, there is still no serious difficulty, though in this case we may want to require that $\mu_n(X_n) = 1$ for all n (each μ_n is a *probability measure*). This ensures that the infinite product $\prod \mu_n(A_n)$ is well-defined for any sequence of measurable sets A_n, so we can use this formula to define $\mu(\prod A_n)$.

We now prove a series of results relating product measures to iterated integrals. We need the following notation. Let X and Y be sets, let $C \subseteq$

$X \times Y$, and let $f : X \times Y \to \mathbf{F}$ be a function. Then for $x \in X$ and $y \in Y$ we define sets $C_x \subseteq Y$ and $C^y \subseteq X$ by

$$C_x = \{y' \in Y : (x, y') \in C\} \qquad \text{and} \qquad C^y = \{x' \in X : (x', y) \in C\}$$

and functions $f_x : Y \to \mathbf{F}$ and $f^y : X \to \mathbf{F}$ by

$$f_x(y') = f(x, y') \qquad \text{and} \qquad f^y(x') = f(x', y).$$

These are called *sections* of C and f.

The first result in this series contains all of the technical content. It states that the measure of a subset of a product can be found by integrating the measures of its sections. One slight subtlety is that the sections of a measurable subset of a product space might not all be measurable subsets of the factor spaces. However, almost all sections are. So in the following theorem the functions $x \mapsto \nu(C_x)$ and $y \mapsto \mu(C^y)$ are in general only defined almost everywhere, but of course that is good enough for them to have well-defined integrals.

Theorem 2.34. *(Cavalieri's principle) Let (X, μ) and (Y, ν) be complete σ-finite measure spaces and let C be a measurable subset of the product measure space. Then $C_x \subseteq Y$ and $C_y \subseteq X$ are measurable for almost every x and y, the functions $x \mapsto \nu(C_x)$ and $y \mapsto \mu(C^y)$ are measurable, and*

$$(\mu \times \nu)(C) = \int \nu(C_x) \, d\mu = \int \mu(C^y) \, d\nu.$$

Proof. The theorem is trivial when C is a finite measurable rectangle, and it then follows from Theorem 2.6 (d) and Theorem 2.27 (c) for countable unions of disjoint finite measurable rectangles. (The part about C_x and C^y being measurable actually holds for all x and y.)

Now assume $\mu(X), \nu(Y) < \infty$ and suppose $C = \bigcap C_n$ where each C_n is a countable union of disjoint finite measurable rectangles. Since the intersection of two countable unions of disjoint finite measurable rectangles is itself a countable union of disjoint finite measurable rectangles, we can assume that $C_1 \supseteq C_2 \supseteq \cdots$. The theorem then holds for C by Theorems 2.6 (d), 2.9 (d), and DCT. We still have C_x and C^y measurable for all x and y.

If $C \subseteq X \times Y$ is an arbitrary measurable subset then by the construction of Theorem 2.14 we can find a sequence of sets, each of which is a countable union of finite measurable rectangles — without loss of generality a countable disjoint union — which contains C and whose measure converges to $(\mu \times \nu)(C)$. Intersecting them yields a set C' for which the theorem holds, which contains C, and such that $C' \setminus C$ is null. Applying the same argument to the complement of C yields another set C'' for which the theorem holds, which is contained in C, and such that $C \setminus C''$ is null. Then letting $f(x) = \nu(C'_x)$ and $g(x) = \nu(C''_x)$, we have that $g \leq f$, both functions are measurable, and

$$(\mu \times \nu)(C) = \int f \, d\mu = \int g \, d\mu.$$

This implies that $f = g$ almost everywhere, i.e., $\nu(C'_x) = \nu(C''_x)$ for almost every x, and since $C''_x \subseteq C_x \subseteq C'_x$ for all x, we conclude (using the fact that (Y, ν) is complete) that C_x is measurable for almost every x and that $(\mu \times \nu)(C) = \int \nu(C_x) \, d\mu$. A similar argument leads to the same conclusion for the y sections. This completes the proof in the finite measure case.

Passing to the σ-finite case, if X and Y are both σ-finite then we can express $X \times Y$ as a countable union of disjoint finite measurable rectangles. For any measurable subset C of $X \times Y$, the preceding argument can be applied to the intersection of C with each of these rectangles, and then taking a countable sum yields the desired result.   □

Cavalieri's principle can be modified to apply not just to subsets of the product space, but to measurable functions. This modification comes in two versions: one (Tonelli) applies to arbitrary positive measurable functions and the other (Fubini) applies to integrable scalar-valued functions. In a typical application, given a scalar-valued function f one might first apply Tonelli to $|f|$ in order to show that its integral is finite, and then, having shown that f is integrable, apply Fubini to f.

We can make the same comment for Tonelli and Fubini that we made for Cavalieri, about the sections f_x and f^y only being measurable for almost every x and y but this being good enough to allow them to be integrated.

Theorem 2.35. *(Tonelli's theorem) Let (X, μ) and (Y, ν) be complete σ-finite measure spaces and let $f : X \times Y \to [0, \infty]$ be measurable. Then f_x and f^y are measurable functions for almost every x and y, the functions $x \mapsto \int f_x \, d\nu$ and $y \mapsto \int f^y \, d\mu$ are measurable, and*

$$\int f \, d(\mu \times \nu) = \int \left[\int f(x, y) \, d\nu \right] d\mu = \int \left[\int f(x, y) \, d\mu \right] d\nu.$$

Proof. The result holds when f is a characteristic function by Theorem 2.34, and it follows for simple functions by linearity. For a general measurable function f, find an increasing sequence of simple functions h_n which converges pointwise to f as in the comment preceding Theorem 2.27. Then

MCT implies that $\int (h_n)_x \, d\nu \to \int f_x \, d\nu$ and $\int (h_n)^y \, d\mu \to \int f^y \, d\mu$ for all x and y off of null sets on which $(h_n)_x$ and $(h_n)^y$ are not measurable, so measurability of the functions $x \mapsto \int (h_n)_x \, d\nu$ and $y \mapsto \int (h_n)^y \, d\mu$ implies measurability of the functions $x \mapsto \int f_x \, d\nu$ and $y \mapsto \int f^y \, d\mu$. Applying MCT throughout the equations

$$\int h_n \, d(\mu \times \nu) = \int \left[\int h_n(x,y) \, d\nu \right] d\mu = \int \left[\int h_n(x,y) \, d\mu \right] d\nu$$

(and using the fact that the inner integrals increase to $\int f_x \, d\nu$ and $\int f^y \, d\mu$, as shown above) yields the desired equality of integrals. $\qquad\square$

Theorem 2.36. *(Fubini's theorem) Let (X,μ) and (Y,ν) be complete σ-finite measure spaces and let $f : X \times Y \to \mathbf{F}$ be integrable. Then f_x and f^y are integrable for almost every x and y, respectively, the functions $x \mapsto \int f_x \, d\nu$ and $y \mapsto \int f^y \, d\mu$ are integrable, and*

$$\int f \, d(\mu \times \nu) = \int \left[\int f(x,y) \, d\nu \right] d\mu = \int \left[\int f(x,y) \, d\mu \right] d\nu.$$

Proof. Since f is integrable, i.e., $\int |f| < \infty$, Tonelli's theorem implies that $\int |f|_x \, d\nu$ and $\int |f|^y \, d\mu$ are finite almost everywhere. (Otherwise the iterated integral in Tonelli's theorem would be infinite, since a function which is infinite on a set of positive measure must have an infinite integral.) This shows that f_x and f^y are integrable for almost every x and y, respectively. The remainder of the theorem follows from Tonelli's theorem by writing f as a linear combination of positive integrable functions. $\qquad\square$

2.7 Scalar-valued measures

Measures can be generalized to allow them to take negative or complex values. One motivation for doing this is suggested by part (d) of Theorem 2.27, where we created a new measure by integrating a positive measurable function f. The same construction makes sense for any integrable function, with the result now being something that looks like a measure but which may take negative or complex values.

A second motivation is the fact that the sum of two measures defined on the same σ-algebra will again be a measure, as will the product of a measure by a positive scalar. This suggests that the positive measures are really just the "positive part" of a vector space of scalar-valued measures. This interpretation will be pursued further in Section 4.5.

Definition 2.37. Let X be a measurable space with σ-algebra Ω. A *scalar-valued measure* on X is a function $\mu : \Omega \to \mathbf{F}$ such that $\mu(\emptyset) = 0$ and $\sum \mu(A_n)$ converges absolutely to $\mu(\bigcup A_n)$ whenever (A_n) is a sequence of disjoint measurable sets. A scalar-valued measure is a *complex measure* if $\mathbf{F} = \mathbf{C}$ and a *signed measure* if $\mathbf{F} = \mathbf{R}$.

Some authors allow signed measures to take the value ∞ or $-\infty$ (but not both), but we will not need to do this.

In practice, one often leaves off the qualifier and simply speaks of "measures" if the context makes it clear whether one means positive, signed, or complex measures.

Scalar-valued measures are not significantly more general than positive measures, in the sense that every scalar-valued measure is a linear combination of positive measures. In the signed case this statement can even be slightly sharpened to say that every signed measure is the difference of two positive measures whose supports are disjoint, in the following sense.

We write $\mu|_A$ for the restriction of a measure μ to the measurable subsets of A.

Definition 2.38. A measurable set A is *null* for a scalar-valued measure μ if $\mu|_A = 0$, i.e., $\mu(B) = 0$ for all measurable $B \subseteq A$. Two measures μ and ν defined on the same σ-algebra are *mutually singular*, written $\mu \perp \nu$, if there exists a measurable set A such that A is null for μ and A^c is null for ν.

Lemma 2.39. *Every scalar-valued measure is bounded. If μ is a signed measure and C is a measurable set such that $\mu(C) > 0$ then there exists a measurable subset $B \subseteq C$ such that $\mu(B) \geq \mu(C)$ and $\mu|_B$ is positive.*

Proof. Let μ be a scalar-valued measure on a set X and suppose X has subsets of arbitrarily large measure in absolute value. We construct a descending sequence of subsets $A_1 \supset A_2 \supset \cdots$ such that each A_n has subsets of arbitrarily large measure in absolute value and $|\mu(A_n \setminus A_{n+1})| \geq 1$ for all n. Start by letting $A_1 = X$. Having constructed A_n, since it has subsets of arbitrarily large measure we can find $B \subset A_n$ such that $|\mu(B)| \geq |\mu(A_n)| + 1$. Then $\mu(A_n) = \mu(B) + \mu(A_n \setminus B)$ implies that $|\mu(A_n \setminus B)| \geq 1$. Also, since A_n has subsets of arbitrarily large measure the same must be true of either B or $A_n \setminus B$. Let A_{n+1} be either B or $A_n \setminus B$ accordingly; this completes the construction.

The sequence of differences $(A_n \setminus A_{n+1})$ now violates the countable additivity condition on scalar-valued measures, because the series $\sum \mu(A_n \setminus$

A_{n+1}) cannot converge absolutely — the terms do not even go to zero. We conclude that every scalar-valued measure must be bounded.

For the second assertion, let μ be a signed measure, suppose $\mu(C) > 0$, and define a sequence (C_n) of disjoint subsets of C as follows. Let $C_1 = \emptyset$. Inductively, let

$$s_n = \inf\{\mu(B) : B \subseteq C \setminus (C_1 \cup \cdots \cup C_n)\}$$

and choose $C_{n+1} \subseteq C \setminus (C_1 \cup \cdots \cup C_n)$ such that $\mu(C_{n+1}) \le \min\left(0, s_n + \frac{1}{n}\right)$. Let $B = C \setminus \bigcup C_n$. Then

$$\mu(B) = \mu(C) - \sum \mu(C_n) \ge \mu(C)$$

since $\mu(C_n) \le 0$ for all n. For any $B' \subseteq B$, if $\mu(B') < 0$ then $\mu(B') < -\frac{1}{n}$ for some n, and then $\mu(C_{n+1} \cup B') < \mu(C_{n+1}) - \frac{1}{n} \le s_n$, contradicting the definition of s_n. We conclude that $\mu|_B$ is positive. $\qquad\square$

Theorem 2.40. *(Hahn-Jordan decomposition theorem) Let μ be a signed measure on a set X.*

(a) *There exist complementary measurable sets $P, N \subseteq X$ such that $\mu|_P$ and $-\mu|_N$ are both finite positive measures.*
(b) *There is a unique decomposition $\mu = \mu^+ - \mu^-$ such that μ^+ and μ^- are mutually singular finite positive measures.*

Proof. (a) Let $a = \sup\{\mu(A) : A \text{ is measurable}\} < \infty$. We claim that there is a measurable set P such that $\mu|_P$ is positive and $\mu(P) = a$. Given this, let $N = X \setminus P$; then $-\mu|_N$ must be positive, since if $\mu(B) > 0$ for some $B \subseteq N$ then $\mu(P \cup B) = a + \mu(B) > a$, a contradiction. Finiteness of $\mu|_P$ and $\mu|_N$ follows from boundedness of μ.

To prove the claim, let (A_n) be a sequence of measurable sets such that $\mu(A_n) \to a$. For each n use the lemma to find a subset B_n of A_n such that $\mu(B_n) \ge \mu(A_n)$ and $\mu|_{B_n}$ is positive. (If $\mu(A_n) \le 0$ then we can take $B_n = \emptyset$.) Then $P = \bigcup B_n$ has the desired properties.

(b) Let P and N be as in (a) and define $\mu^+(A) = \mu(A \cap P)$ and $\mu^-(A) = -\mu(A \cap N)$ for all measurable A. It is clear that μ^+ and μ^- are mutually singular positive measures such that $\mu = \mu^+ - \mu^-$. If ν^+, ν^- are any other such pair, let P' and N' be complementary measurable sets witnessing the mutual singularity of ν^+ and ν^-. Then $P \setminus P' = P \cap N'$ implies that $\mu|_{P \setminus P'} = 0$, and similarly for $P' \setminus P$. Thus

$$\mu^+(A) = \mu(A \cap P) = \mu(A \cap P \cap P') = \mu(A \cap P') = \nu^+(A)$$

for all measurable $A \subseteq X$, so that $\mu^+ = \nu^+$, and similarly $\mu^- = \nu^-$. $\qquad\square$

Thus, every signed measure is the difference of two positive measures with disjoint supports. The version given in part (a) is perhaps more intuitive, but it is not canonical because the sets P and N could be modified on null sets. (This is the only ambiguity, however, so the result is canonical up to null sets, which is almost as good.)

Since the real and imaginary parts of any complex measure are signed measures, it immediately follows that every complex measure is a linear combination of four positive measures. However, the supports of the decompositions of the real and imaginary parts could overlap. There is a cleaner reduction of complex measures to positive measures based on the polar decomposition of a complex number, but before presenting it we first need to describe a positive measure that plays the role of the absolute value of a complex number. In the signed case this would just be $\mu^+ + \mu^-$, but for complex measures we need a different definition.

If $A = \bigcup_{n=1}^{\infty} A_n$ where the A_n are measurable and disjoint, then we say that the A_n consititute a *measurable partition* of A.

Proposition 2.41. *Let μ be a scalar-valued measure. For any measurable set A, define*

$$|\mu|(A) = \sup\left\{\sum_{n=1}^{\infty} |\mu(A_n)| : (A_n) \text{ is a measurable partition of } A\right\}.$$

Then

(i) *$|\mu|$ is a positive measure;*
(ii) *$|\mu|$ is the smallest positive measure which satisfies $|\mu(A)| \leq |\mu|(A)$ for all measurable A;*
(iii) *if μ is a signed measure then $|\mu| = \mu^+ + \mu^-$;*
(iv) *$|\mu|$ is finite.*

Proof. (i) $|\mu|(\emptyset) = 0$ is clear. For countable additivity, let (A_n) be a sequence of disjoint measurable sets and let $A = \bigcup A_n$. Given measurable partitions of each A_n, we can amalgamate them into a measurable partition of A, and this gives us $|\mu|(A) \geq \sum |\mu|(A_n)$. Conversely, any measurable partition of A can be intersected with each A_n, which gives us $|\mu|(A) \leq \sum |\mu|(A_n)$. So $|\mu|$ is countably additive.

(ii) It is clear that $|\mu(A)| \leq |\mu|(A)$ for all A. Suppose ν is any positive measure which satisfies $|\mu(A)| \leq \nu(A)$ for all A. Then for any A we have

$$|\mu|(A) = \sup\left\{\sum |\mu(A_n)| : (A_n) \text{ is a measurable partition of } A\right\}$$

$$\leq \sup \left\{ \sum \nu(A_n) : (A_n) \text{ is a measurable partition of } A \right\} = \nu(A).$$

Thus $|\mu|$ is the smallest positive measure with this property.

(iii) Suppose μ is a signed measure and let P and N be as in Theorem 2.40. Let $\nu = \mu^+ + \mu^-$. Then for any A we have

$$|\mu|(A) \geq |\mu(A \cap P)| + |\mu(A \cap N)| = \nu(A).$$

However, ν is a positive measure which satisfies $|\mu(A)| \leq \nu(A)$ for all A, so part (ii) shows that $|\mu| \leq \nu$. Hence $|\mu| = \nu = \mu^+ + \mu^-$.

(iv) If μ is a complex measure we can write $\mu = \mu_1 + i\mu_2$ where μ_1 and μ_2 are signed measures, and it follows from part (ii) that $|\mu| \leq |\mu_1| + |\mu_2|$. This reduces to the signed case, which follows from part (iii) and the fact that μ^+ and μ^- are finite in Theorem 2.40. $\quad\square$

Note that in the definition of $|\mu|$ we can restrict to partitions of A into finitely many measurable pieces. This is because $\sum_{n=1}^{\infty} |\mu(A_n)|$ is the limit of its partial sums, so the contribution of the infinite measurable partition (A_n) is approximated by the contribution of the finite partitions $\{A_1, \ldots, A_N, \bigcup_{n=N+1}^{\infty} A_n\}$.

The measure $|\mu|$ is called the *variation* of μ. We will show that any complex measure can be expressed as the integral against its variation of a function taking values in the unit circle. This is the canonical reduction of complex to positive measures.

We will achieve this result as a byproduct of a general analysis of the relation between two measures defined on the same σ-algebra. We need the following notion.

Definition 2.42. Let μ be a positive measure and let ν be a scalar-valued measure defined on the same σ-algebra. Then ν is *absolutely continuous* with respect to μ, written $\nu \ll \mu$, if $\mu(A) = 0$ implies $\nu(A) = 0$, for all measurable A.

This condition on null sets has a surprisingly strong consequence. If $\nu(A) = 0$ whenever $\mu(A) = 0$, then $|\nu(A)|$ must also be small whenever $\mu(A)$ is small, in the following sense.

Proposition 2.43. *Let μ be a positive measure and let ν be a scalar-valued measure defined on the same σ-algebra. Then $\nu \ll \mu$ iff for every $\epsilon > 0$ there exists $\delta > 0$ such that $\mu(A) < \delta$ implies $|\nu(A)| < \epsilon$.*

Proof. The reverse direction is easy: if $\mu(A) = 0$ then $\mu(A) < \delta$ for all δ, hence $|\nu(A)| < \epsilon$ for all ϵ, hence $\nu(A) = 0$. For the forward direction, suppose the ϵ-δ condition fails. Then there exists a value of ϵ such that the condition fails for all δ. Thus, for each n there is a measurable set A_n such that $\mu(A_n) < 2^{-n}$ but $|\nu(A_n)| \geq \epsilon$. Let

$$A = \limsup A_n = \bigcap_{k=1}^{\infty} \bigcup_{i=k}^{\infty} A_i;$$

then $\mu(A) \leq \mu(\bigcup_{i=k}^{\infty} A_i) < 2^{-(k-1)}$ for all k, and hence $\mu(A) = 0$. However, $|\nu|(\bigcup_{i=k}^{\infty} A_i) \geq \epsilon$ for all k, which implies $|\nu|(A) \geq \epsilon$ (using Theorem 2.9 (d) and the fact that $|\nu|$ is finite). From the definition of $|\nu|$ it follows that there is a measurable set $B \subseteq A$ such that $\nu(B) \neq 0$, and we have contradicted the assumption of absolute continuity. This completes the proof. $\qquad\square$

Absolute continuity is at the opposite extreme to mutual singularity. Our first basic comparison result states that in combination these two cases cover all the possibilities.

Theorem 2.44. *(Lebesgue decomposition theorem) Let μ be a positive measure and let ν be a scalar-valued measure defined on the same σ-algebra. Then there are scalar-valued measures ν_1 and ν_2 such that $\nu_1 \ll \mu$, $\nu_2 \perp \mu$, and $\nu = \nu_1 + \nu_2$. This decomposition of ν is unique.*

Proof. By Proposition 2.41 (iv) $|\nu|$ is finite. Thus $a = \sup\{|\nu|(A) : \mu(A) = 0\} < \infty$. Find a sequence of measurable sets (A_n) such that $\mu(A_n) = 0$ for all n and $|\nu|(A_n) \to a$ and let $A = \bigcup A_n$. Then $\mu(A) = 0$ and $|\nu|(A) = a$.

Let ν_1 be the restriction of ν to A^c and let ν_2 be the restriction of ν to A. It is clear that $\nu = \nu_1 + \nu_2$ and that $\nu_2 \perp \mu$. If ν_1 failed to be absolutely continuous with respect to μ then there would be a set B such that $\mu(B) = 0$ but $\nu_1(B) \neq 0$. Without loss of generality $B \cap A = \emptyset$, and $|\nu|(A \cup B) = |\nu|(A) + |\nu|(B) > a$ then contradicts the definition of a. Thus $\nu_1 \ll \mu$.

Let $\nu = \nu_1' + \nu_2'$ be any other decomposition of ν satisfying $\nu_1' \ll \mu$ and $\nu_2' \perp \mu$. Let B be a measurable set such that $\mu(B) = 0$ and ν_2' is supported on B. Then ν_1 and ν_1' are both supported on the complement of $A \cup B$ since μ is zero on $A \cup B$, which implies that $\nu_2 = \nu|_{A \cup B} = \nu_2'$. Thus $\nu_2 = \nu_2'$ and therefore also $\nu_1 = \nu_1'$, proving that the decomposition is unique. $\qquad\square$

It follows from Theorem 2.27 (d) that if μ is a positive measure then for any μ-integrable scalar-valued function f the map $A \mapsto \int_A f \, d\mu$ defines a scalar-valued measure. (Write f as a linear combination of positive integrable functions.) Any measure of this form is clearly absolutely continuous with respect to μ. The following theorem shows that if μ is σ-finite then the converse is true, i.e., every $\nu \ll \mu$ arises in this way.

In our proof of Cavalieri's principle (Theorem 2.34) we used a standard technique for dealing with σ-finite spaces: if X is σ-finite we can write it as the union of a sequence of disjoint finite measure subsets A_n, and work on each A_n separately. In the following proof we introduce a different technique. Given such a decomposition, we may assume that $\mu(A_n) > 0$ for all n and define

$$\tilde{\mu}(A) = \sum \frac{1}{2^n \mu(A_n)} \mu(A \cap A_n).$$

Since we have $\mu(A) = \sum \mu(A \cap A_n)$ by countable additivity, all we have done is to have scaled down the measure by a uniform amount on each A_n. Then $\tilde{\mu}$ is a finite measure, and it is easy to see that

$$\int f \, d\mu = \int f w \, d\tilde{\mu}$$

for all positive measurable functions f, where w is the *weight function* which takes the value $2^n \mu(A_n)$ on A_n. (Check this first when f is a characteristic function, then appeal to linearity and MCT.) The same formula holds for scalar-valued functions f, with the additional comment that f is μ-integrable if and only if $f w$ is $\tilde{\mu}$-integrable.

Theorem 2.45. *(Radon-Nikodym theorem) Let (X, μ) be a σ-finite measure space and let ν be a scalar-valued measure which is absolutely continuous with respect to μ. Then there exists a μ-integrable scalar-valued function f such that $\nu(A) = \int_A f \, d\mu$ for all measurable sets A. Moreover, $|\nu|(A) = \int_A |f| \, d\mu$ for all A.*

Proof. By applying the Hahn-Jordan decomposition to the real and imaginary parts of ν, we can write ν as a linear combination of four finite positive measures, each of which is absolutely continuous with respect to μ. By treating these measures separately we can reduce the first part of the theorem to the case where ν is a finite positive measure.

Suppose first that μ is also finite. Fix $n \in \mathbf{N}$ and for each $k \in \mathbf{N}$ let N_k be the support of $(\nu - \frac{k}{2^n} \mu)^-$ (the negative part of the signed measure $\nu - \frac{k}{2^n} \mu$, as in Theorem 2.40). Observe that $N_1 \subseteq N_2 \subseteq \cdots$. Then set $N_0 = \emptyset$ and define $f_n = \sum_{k=1}^\infty \frac{k}{2^n} \cdot 1_{N_{k+1} \setminus N_k}$ for $n \geq 1$. For any subset A of $N_{k+1} \setminus N_k$ we have

$$\int_A f_n \, d\mu = \frac{k}{2^n} \mu(A) \leq \nu(A) \leq \frac{k+1}{2^n} \mu(A) = \int_A \left(f_n + \frac{1}{2^n} \right) d\mu.$$

Also, for $B = (\bigcup N_k)^c$ we have $\nu(B) \geq \frac{k}{2^n}\mu(B)$ for all k, which implies that $\mu(B) = 0$, and hence $\nu(B) = 0$ also since $\nu \ll \mu$. So for any measurable set A, summing over $A \cap (N_k \setminus N_{k-1})$ yields

$$\int_A f_n \, d\mu \leq \nu(A) \leq \int_A f_n \, d\mu + \frac{\mu(X)}{2^n}.$$

Now we just have to observe that the sequence (f_n) is increasing and let $f = \lim f_n$; by MCT the preceding implies that $\nu(A) = \int_A f \, d\mu$ for all measurable sets A.

We now have the first part of the theorem, assuming μ is finite. For the second part define $\nu'(A) = \int_A |f| \, d\mu$; this is a positive measure by Theorem 2.27 (d). For any measurable set A we have

$$|\nu(A)| = \left| \int_A f \, d\mu \right| \leq \int_A |f| \, d\mu = \nu'(A).$$

By Proposition 2.41 (ii) it follows that $|\nu| \leq \nu'$. Conversely, given any measurable set A and any simple function h defined on A with standard form $h = \sum a_i 1_{A_i}$ and such that each $|a_i| = 1$, we have

$$|\nu|(A) \geq \sum |\nu(A_i)| \geq \left| \int_A h \cdot f \, d\mu \right|;$$

applying this to a sequence of simple functions of this form which converge pointwise to $\frac{|f|}{f}$ and invoking DCT then yields

$$\nu'(A) = \int_A |f| \, d\mu = \int_A \frac{|f|}{f} \cdot f \, d\mu \leq |\nu|(A).$$

So we conclude that $|\nu| = \nu'$.

Finally, suppose μ is only σ-finite and define $\tilde{\mu}$ and w as in the comment preceding the theorem. Then ν is absolutely continuous with respect to $\tilde{\mu}$, so we can find a $\tilde{\mu}$-integrable function f such that $\nu(A) = \int_A f \, d\tilde{\mu}$ for all measurable sets A. It immediately follows that fw^{-1} is μ-integrable and $\nu(A) = \int_A fw^{-1} \, d\mu$ for all measurable sets A. We similarly have

$$|\nu|(A) = \int_A |f| \, d\tilde{\mu} = \int_A |fw^{-1}| \, d\mu$$

for all A.   □

This yields the promised reduction of complex measures to positive measures.

Corollary 2.46. *Let ν be a complex measure on a measurable space X. Then there is a finite positive measure $|\nu|$ and a measurable function f : $X \to \mathbf{C}$ with $|f(x)| = 1$ $|\nu|$-almost everywhere, such that*

$$\nu(A) = \int_A f \, d|\nu|$$

for any measurable set A.

This follows from Theorem 2.45 by taking $\mu = |\nu|$. In this case the second part of that result says that $|\nu|(A) = \int_A |f| \, d|\nu|$ for every measurable set A; since we also have $|\nu|(A) = \int_A 1 \, d|\nu|$ for all A, this entails that $|f| = 1$ almost everywhere.

One sometimes writes $d\nu = f \, d\mu$ or even $f = \frac{d\nu}{d\mu}$ to indicate the relationship $\nu(A) = \int_A f \, d\mu$ in the Radon-Nikodym theorem. The function f is called the *Radon-Nikodym derivative* of ν with respect to μ. Like the usual derivative, it satisfies a chain rule. If f is $|\mu|$-integrable, define $\int f \, d\mu = \int f \, d\mu_1 - \int f \, d\mu_2 + i \int f \, d\mu_3 - i \int f, d\mu_4$ where $\mu = \mu_1 - \mu_2 + i\mu_3 - i\mu_4$ and the μ_i are positive.

Proposition 2.47. *Let μ be a σ-finite measure, let ν be a scalar-valued measure, and suppose $\nu \ll \mu$. Then for any ν-integrable function g the function $g \cdot \frac{d\nu}{d\mu}$ is μ-integrable and satisfies*

$$\int g \, d\nu = \int g \cdot \frac{d\nu}{d\mu} \, d\mu.$$

Proof. We can reduce to the case that ν is positive in the usual way. The conclusion is true when $g = 1_A$ is a characteristic function by the definition of $\frac{d\nu}{d\mu}$. It is then true for simple functions by linearity, for positive integrable functions by MCT, and for all integrable functions by linearity again. $\square$

In particular, it follows that

$$\int f \, d\nu = \int f \cdot \frac{d\nu}{d|\nu|} \, d|\nu|.$$

This can be taken as an alternative definition of the integral against a scalar-valued measure.

2.8 Exercises

2.1. Let Ω be a family of subsets of a set that includes X and is stable under differences and countable disjoint unions. Prove that Ω is a σ-algebra.

2.2. Prove the observation that was made just before Lemma 2.4.

2.3. Prove that a function $f : X \to \mathbf{R}$ is measurable if and only if $f^{-1}([a, \infty))$ is a measurable subset of X for all $a \in \mathbf{Q}$.

2.4. Let X be a measurable space and let (f_n) be a sequence of measurable functions into $\mathbf{F}$. Show that $\{x \in X : \lim f_n(x) \text{ exists}\}$ is measurable.

2.5. What is wrong with the claim that the condition $\mu(\emptyset) = 0$ in Definition 2.7 is redundant because countable additivity entails that $\mu(\emptyset) = \sum_{n=1}^{\infty} \mu(\emptyset)$ and hence that $\mu(\emptyset) = 0$.

2.6. Let (X, μ) be a σ-finite measure space and let $A \subseteq X$ be measurable with $\mu(A) = \infty$. Show that for any $N \in \mathbf{N}$ there exists a measurable set $B \subseteq A$ such that $N \leq \mu(B) < \infty$.

2.7. Let μ be a complete measure on a set X and suppose $f_n \to f$ a.e., with each f_n measurable. Show that f is measurable.

2.8. (Egoroff's theorem) Let (X, μ) be a finite measure space, let $f_n : X \to \mathbf{F}$ be measurable, and suppose $f_n \to f$ pointwise. Prove that for every $\epsilon > 0$ there exists $A \subseteq X$ such that $\mu(A) \leq \epsilon$ and $f_n \to f$ uniformly on A^c. (First prove that for any $\delta > 0$ there exists a measurable set $B \subseteq X$ such that $\mu(B) \leq \delta$ and $|f - f_n| < \delta$ on B^c for sufficiently large n.)

2.9. In Theorem 2.14, let μ' be any complete measure whose domain contains Ω_0 and which agrees with μ_0 on Ω_0. Prove that the domain of μ' contains Ω and μ' agrees with μ on Ω.

2.10. In the proof of Theorem 2.14, assume $\mu_0(X) < \infty$ and define the *inner measure* of any subset $A \subseteq X$ by $\mu_*(A) = \mu_0(X) - \mu^*(A^c)$. Show that A belongs to Ω if and only if $\mu^*(A) = \mu_*(A)$.

2.11. Evaluate the Lebesgue measure of the Cantor set K. Modify the middle thirds construction of the Cantor set by removing shorter intervals to produce a set which is homeomorphic to K but whose measure is any desired value in $(0, 1)$.

2.12. Let $F : \mathbf{R} \to \mathbf{R}$ satisfy (1) $s < t$ implies $F(s) \leq F(t)$ and (2) if s_n decreases to s then $F(s_n) \to F(s)$. Show that there is a measure μ_F on $\mathbf{R}$ whose domain contains every open set and such that $\mu_F((a, b]) = F(b) - F(a)$ for all $a, b \in \mathbf{R}$, $a < b$.

2.13. Use Exercise 2.12 to find a nonzero positive measure μ on $\mathbf{R}$ whose domain contains every open set, such that (1) there is an open set U with $\mu(U) = m(U^c) = 0$ and (2) we have $\mu(\{x\}) = 0$ for all $x \in \mathbf{R}$.

2.14. Let (X, μ) be a measure space and let $f, f_n : X \to [0, \infty]$ be measurable functions. Suppose $f_n \to f$ pointwise and $\lim \int f_n = \int f < \infty$. Show that $\int_A f_n \to \int_A f$ for all measurable $A \subseteq X$. Give an example where this is not true if $\lim \int f_n = \int f = \infty$.

2.15. Let (X, μ) be a measure space and let $f : X \to [0, \infty]$ be measurable. Suppose $\int f < \infty$. Given $\epsilon > 0$, show that there is a measurable set $A \subseteq X$ such that $\mu(A) < \infty$ and $\int_A f \geq \int f - \epsilon$.

2.16. Let f and g be integrable. Show that $\int_A f = \int_A g$ for all measurable A iff $\int |f - g| = 0$ iff $f = g$ almost everywhere.

2.17. Let f and g be integrable real-valued functions on a complete measure space and assume $f \leq h \leq g$ and $\int f = \int g$. Prove that h is measurable.

2.18. If (X, μ) is any measure space and $\overline{\mu}$ is the completion of μ, show that every $\overline{\mu}$-measurable function $f : X \to \mathbf{F}$ almost everywhere equals a μ-measurable function $g : X \to \mathbf{F}$. (Approximate by simple functions.)

2.19. Let (f_n) be a sequence of measurable functions on a measure space X. Suppose that $\sum \int |f_n|$ is finite. Use product measures to prove that $\sum \int f_n = \int \sum f_n$.

2.20. Show that the product measure on the product of n copies of $\mathbf{R}$ equipped with Lebesgue measure equals Lebesgue measure on $\mathbf{R}^n$ (including the fact that the two measures are defined on the same σ-algebra).

2.21. Let ν be a complex measure on a set X and suppose $\nu(X) = |\nu|(X)$. Prove that ν is positive.

2.22. Let μ be a positive measure and let f be an integrable real-valued function. Prove that for every $\epsilon > 0$ there exists $\delta > 0$ such that $\mu(A) < \delta$ implies $|\int_A f \, d\mu| < \epsilon$.

Chapter 3

Banach Spaces

3.1 Normed vector spaces

One of the reasons why group theory is so important is because groups often appear as auxiliaries to other kinds of mathematical objects. The permutations of any set form a group, and if we restrict to those permutations which leave unchanged some additional structure carried by the set, the result should still be a group. In this way we get automorphism groups, self-homeomorphism groups, and so on.

A similar story can be told about vector spaces. The functions from any set into a given field form a vector space over that field, and here too the functions which respect some additional structure on the set will often still form a vector space. Two constructions of this type, the spaces $C(X)$ and $L^\infty(X)$, will play major roles in this book. The dual of a Banach space can also be considered an example of this phenomenon. Other examples include coordinate rings, spaces of analytic functions, and Lipschitz algebras.

When vector spaces arise in this way they typically come equipped with metric or topological structure which is, in a natural sense, compatible with the algebraic operations. Topological considerations are less crucial in finite dimensions because in this setting every vector space over $\mathbf{R}$ or $\mathbf{C}$ (the two cases of interest to us here) is isomorphic to $\mathbf{R}^n$ or $\mathbf{C}^n$ and there is only one reasonable topology, with respect to which every linear subspace is automatically closed and every linear map is automatically continuous. But none of these statements remains true in infinite dimensions, and here it becomes much more important to take topological information into account.

Recall that $\mathbf{F}$ denotes either $\mathbf{R}$ or $\mathbf{C}$. We also accordingly let $\mathbf{F}_0$ denote either $\mathbf{Q}$ or $\mathbf{Q} + i\mathbf{Q}$, so that $\mathbf{F}_0$ is a countable dense subfield of $\mathbf{F}$.

Definition 3.1. A *norm* on a vector space E over $\mathbf{F}$ is a map $\|\cdot\| : E \to$

$[0, \infty)$ satisfying

  (i) $\|x\| = 0$ iff $x = 0$;

  (ii) $\|ax\| = |a|\|x\|$;

  (iii) $\|x + y\| \leq \|x\| + \|y\|$

for all $x, y \in E$ and all $a \in \mathbf{F}$. When E is equipped with a norm we call it a *normed vector space*. A *Banach space* is a normed vector space for which the associated metric $d(x, y) = \|x - y\|$ is complete.

Unit balls are referenced so frequently that they deserve a special notation. We write $(E)_1$ for the *open unit ball*

$$(E)_1 = \{x \in E : \|x\| < 1\}$$

and $[E]_1$ for the *closed unit ball*

$$[E]_1 = \{x \in E : \|x\| \leq 1\}$$

of a normed vector space E. More generally, we write $(E)_r$ and $[E]_r$ for the open and closed balls about the origin of any radius $r > 0$.

The map $x \mapsto \|x\|$ is continuous for the norm topology on E (i.e., the metric topology for the metric associated to the norm). In fact, we have the inequality

$$\big| \|x\| - \|y\| \big| \leq \|x - y\|,$$

an easy consequence of the triangle inequality (property (iii) of Definition 3.1). So the condition stated in Proposition 1.19 is verified with $\delta = \epsilon$.

Completeness is valuable in the setting of normed vector spaces for the same reason it is valuable in the setting of general metric spaces: it gives us the greatest possible freedom to take limits. Since we know that every metric space can be completed, it is natural to ask whether the same is true of normed vector spaces. The answer is yes. Given any normed vector space, we can form its completion as a metric space, and then show that the norm and the algebraic operations extend to this completion so as to make it a Banach space. It is straightforward but slightly tedious to verify this. Instead of going through this argument now we defer the result to Section 3.3, where it can be deduced as a trivial consequence of the basic machinery of duality; see Corollary 3.32.

The most familiar example of a norm is the Euclidean norm in finite dimensions.

Example 3.2. The space $\mathbf{F}^n$ with the *Euclidean norm*

$$\|\mathbf{a}\| = (|a_1|^2 + \cdots + |a_n|^2)^{1/2},$$

where $\mathbf{a} = (a_1, \ldots, a_n)$, is a Banach space.

Although this is a simple example, checking that it has all of the desired properties is not completely trivial. The most interesting issue is the triangle inequality. We give a geometric proof in $\mathbf{R}^n$; this covers the complex case too because $\mathbf{C}^n$ is isometric to $\mathbf{R}^{2n}$. (Alternatively, our proof can be adapted to the complex case using complex inner products.)

Define $\langle \mathbf{a}, \mathbf{b} \rangle = a_1 b_1 + \cdots + a_n b_n$ for $\mathbf{a}, \mathbf{b} \in \mathbf{R}^n$, so that $\|\mathbf{a}\|^2 = \langle \mathbf{a}, \mathbf{a} \rangle$. The geometric intuition is that $\langle \mathbf{a}, \mathbf{b} \rangle = 0$ if $\mathbf{a}$ and $\mathbf{b}$ are perpendicular.

The Pythagorean theorem is now an easy computation: if $\langle \mathbf{a}, \mathbf{b} \rangle = 0$ then

$$\|\mathbf{a} + \mathbf{b}\|^2 = \langle \mathbf{a} + \mathbf{b}, \mathbf{a} + \mathbf{b} \rangle$$
$$= \langle \mathbf{a}, \mathbf{a} \rangle + 2\langle \mathbf{a}, \mathbf{b} \rangle + \langle \mathbf{b}, \mathbf{b} \rangle$$
$$= \|\mathbf{a}\|^2 + \|\mathbf{b}\|^2.$$

In particular, this shows that the hypotenuse of a right triangle, $\mathbf{a} + \mathbf{b}$, is always at least as long as either of the other sides, $\mathbf{a}$ and $\mathbf{b}$.

Now to verify the triangle inequality, let $\mathbf{a}$ and $\mathbf{b}$ be any two vectors in $\mathbf{R}^n$. We can assume $\mathbf{a} + \mathbf{b} \neq 0$ to avoid triviality. The idea of the argument is shown in Figure 3.1. We drop a perpendicular from $\mathbf{a}$ onto the

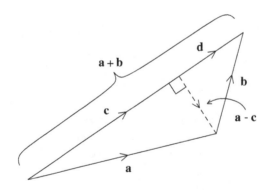

Fig. 3.1   The triangle inequality in $\mathbf{R}^n$

vector $\mathbf{a} + \mathbf{b}$, and this decomposes $\mathbf{a} + \mathbf{b}$ into two pieces, $\mathbf{c}$ and $\mathbf{d}$, which are respectively shorter than $\mathbf{a}$ and $\mathbf{b}$ by the Pythagorean theorem.

In more detail, the proof goes as follows. Dropping a perpendicular from $\mathbf{a}$ onto $\mathbf{a} + \mathbf{b}$ is just a matter of finding the right coefficient λ so that $\mathbf{c} = \lambda(\mathbf{a} + \mathbf{b})$ satisfies $\langle \mathbf{a} - \mathbf{c}, \mathbf{a} + \mathbf{b} \rangle = 0$. Solving for λ yields

$$\lambda = \frac{\langle \mathbf{a}, \mathbf{a} + \mathbf{b} \rangle}{\|\mathbf{a} + \mathbf{b}\|^2},$$

and it is easy to check that indeed $\langle \mathbf{a}, \mathbf{a} + \mathbf{b} \rangle = \langle \mathbf{c}, \mathbf{a} + \mathbf{b} \rangle$ for $\mathbf{c} = \lambda(\mathbf{a} + \mathbf{b})$. We then have $\mathbf{d} = (\mathbf{a} + \mathbf{b}) - \mathbf{c} = \mu(\mathbf{a} + \mathbf{b})$ where

$$\mu = 1 - \lambda = \frac{\langle \mathbf{b}, \mathbf{a} + \mathbf{b} \rangle}{\|\mathbf{a} + \mathbf{b}\|^2}.$$

We now have a right triangle with hypotenuse $\mathbf{a}$ and opposing sides $\mathbf{c}$ and $\mathbf{a} - \mathbf{c}$, and a right triangle with hypotenuse $\mathbf{b}$ and opposing sides $\mathbf{d}$ and $\mathbf{b} - \mathbf{d}$. It follows from the Pythagorean theorem that $\|\mathbf{c}\| \leq \|\mathbf{a}\|$ and $\|\mathbf{d}\| \leq \|\mathbf{b}\|$, and therefore

$$\|\mathbf{a} + \mathbf{b}\| = \lambda \|\mathbf{a} + \mathbf{b}\| + \mu \|\mathbf{a} + \mathbf{b}\| \leq \|\mathbf{c}\| + \|\mathbf{d}\| \leq \|\mathbf{a}\| + \|\mathbf{b}\|$$

since $\|\mathbf{c}\| = |\lambda| \|\mathbf{a} + \mathbf{b}\|$ and $\|\mathbf{d}\| = |\mu| \|\mathbf{a} + \mathbf{b}\|$. (The middle inequality could be strict because $\mathbf{c}$ and $\mathbf{d}$ might point in opposite directions.)

Now let us look at some infinite-dimensional examples.

Example 3.3. Denote by l^1 the set of all scalar sequences $\mathbf{a} = (a_n)$ which satisfy $\sum |a_n| < \infty$. It is equipped with the l^1 *norm* $\|\mathbf{a}\|_1 = \sum |a_n|$.

Proposition 3.4. l^1 *is a Banach space.*

Proof.   It is easy to check that l^1 is a vector space with coordinatewise operations. For instance, the coordinatewise sum of two elements of l^1 is also in l^1 because

$$\sum |a_n + b_n| \leq \sum |a_n| + |b_n| = \sum |a_n| + \sum |b_n|.$$

This computation also shows that $\| \cdot \|_1$ satisfies the triangle inequality. The other properties of a norm are equally easy to verify.

The main point is to verify completeness, so let $(\mathbf{a}^i)$ be a Cauchy sequence of elements of l^1. For each $n \in \mathbf{N}$ we have

$$|a_n^i - a_n^j| \leq \sum_{k=1}^{\infty} |a_k^i - a_k^j| = \|\mathbf{a}^i - \mathbf{a}^j\|_1,$$

so $(\mathbf{a}^i)$ being Cauchy implies that for each n the sequence $(a_n^i)_{i=1}^{\infty} \subset \mathbf{F}$ is Cauchy. Therefore these sequences converge and we can define $a_n = \lim_{i \to \infty} a_n^i$. Then $\mathbf{a} = (a_n)$ is in l^1 because for each m we have

$$\sum_{n=1}^{m} |a_n| = \lim_{i \to \infty} \sum_{n=1}^{m} |a_n^i| \leq \sup_i \|\mathbf{a}^i\|_1,$$

and hence $\sum_{n=1}^{\infty} |a_n| \leq \sup_i \|\mathbf{a}^i\|_1$. This supremum is finite because $(\mathbf{a}^i)$ is Cauchy and $\|\mathbf{a}^i\|_1$ is the l^1 distance from $\mathbf{a}^i$ to the zero element.

Finally, to see that $\mathbf{a}^i \to \mathbf{a}$ in l^1 norm (i.e., in the metric associated to the l^1 norm), let $\epsilon > 0$. Find N such that $i, j \geq N$ implies $\|\mathbf{a}^i - \mathbf{a}^j\|_1 \leq \epsilon$. Then for any $i \geq N$ and any $m \in \mathbf{N}$ we have

$$\sum_{n=1}^{m} |a_n - a_n^i| = \lim_{j \to \infty} \sum_{n=1}^{m} |a_n^j - a_n^i| \leq \epsilon,$$

and taking $m \to \infty$ yields $\|\mathbf{a} - \mathbf{a}^i\|_1 \leq \epsilon$ for all $i \geq N$. This shows that $\mathbf{a}^i \to \mathbf{a}$ in l^1 norm. $\square$

The proof that l^1 is complete is a prototype for other completeness proofs. There are three stages. Given a Cauchy sequence of elements of l^1 we start by identifying the object that ought to be its limit. Next we show that this object belongs to l^1. Finally, we verify that the original sequence does converge to this limit in l^1 norm.

The first stage is easy in the case of l^1. We just observe that a Cauchy sequence in l^1 has to be Cauchy on each coordinate, so we can take the limit in each coordinate separately to obtain the putative limit object. In the second stage the trick is to use the fact that $\sum_{n=1}^{\infty} |a_n| = \lim_{m \to \infty} \sum_{n=1}^{m} |a_n|$. So we can show that $\|\mathbf{a}\|_1$ is finite by putting a bound on $\sum_{n=1}^{m} |a_n|$ that is uniform in m. A similar trick is applied in the last stage.

Example 3.5. Denote by l^2 the set of all scalar sequences $\mathbf{a}$ which satisfy $\sum |a_n|^2 < \infty$. It is equipped with the l^2 *norm* $\|\mathbf{a}\|_2 = (\sum |a_n|^2)^{1/2}$.

Proposition 3.6. *l^2 is a Banach space.*

Proof. The sum of two l^2 sequences is again an l^2 sequence because

$$\sum |a_n + b_n|^2 \leq \sum (|a_n|^2 + 2|a_n||b_n| + |b_n|^2) \leq \sum 2(|a_n|^2 + |b_n|^2)$$

(using the inequality $2st \leq s^2 + t^2$ for $s, t \geq 0$), and it is clear that a scalar multiple of an l^2 sequence is again an l^2 sequence. The assertion that $\| \cdot \|_2$ is a norm can be deduced from Example 3.2 using the fact that

$$\left(\sum_{n=1}^{m} |a_n|^2 \right)^{1/2} \to \left(\sum_{n=1}^{\infty} |a_n|^2 \right)^{1/2}$$

as $m \to \infty$. The proof that l^2 is complete is almost identical to the proof that l^1 is complete, except that we use squared absolute values in place of absolute values. $\square$

The space l^2 is special because its norm comes from an inner product, $\|\mathbf{a}\| = \langle \mathbf{a}, \mathbf{a} \rangle^{1/2}$, where $\langle \mathbf{a}, \mathbf{b} \rangle = \sum a_n b_n$ or $\sum a_n \bar{b}_n$ depending on whether the scalars are real or complex. (Of course, $b_n = \bar{b}_n$ if b_n is real, so the second formula is technically valid in both cases. In the sequel, rather than write two separate formulas in this kind of situation, we will just give the complex version and expect the reader to understand that any complex conjugates are superfluous in the real case.) This sum is absolutely convergent because

$$\left(\sum_n |a_n \bar{b}_n| \right)^2 = \sum_n |a_n|^2 |b_n|^2 + \sum_{m<n} 2|a_m \bar{b}_m a_n \bar{b}_n|$$
$$\leq \sum_n |a_n|^2 |b_n|^2 + \sum_{m<n} (|a_m|^2 |b_n|^2 + |a_n|^2 |b_m|^2)$$
$$= \left(\sum_n |a_n|^2 \right) \left(\sum_n |b_n|^2 \right)$$

(again using $2st \leq s^2 + t^2$, here with $s = |a_m \bar{b}_n|$ and $t = |a_n \bar{b}_m|$). The existence of an inner product makes l^2 a Hilbert space, a topic we will explore further in Chapter 5.

Example 3.7. Denote by l^∞ the set of all bounded scalar sequences, equipped with the *sup norm* $\|\mathbf{a}\|_\infty = \sup |a_n|$. Denote by c_0 the subset of l^∞ consisting of those scalar sequences which converge to 0.

Proposition 3.8. l^∞ *and* c_0 *are Banach spaces.*

Proof. Checking that l^∞ and c_0 are normed vector spaces is routine. To verify that l^∞ is complete, we follow the usual procedure. Let $(\mathbf{a}^i)$ be a Cauchy sequence in l^∞; then it is Cauchy on each coordinate, so we can define $a_n = \lim_{i \to \infty} a_n^i$. We then have that $\mathbf{a}^i \to \mathbf{a}$ uniformly because

$$\|\mathbf{a}^j - \mathbf{a}^i\|_\infty \leq \epsilon \text{ for all } i, j \geq N$$

implies

$$|a_n^j - a_n^i| \leq \epsilon \text{ for all } i, j \geq N \text{ and all } n$$

implies

$$|a_n - a_n^i| \leq \epsilon \text{ for all } i \geq N \text{ and all } n$$

implies

$$\|\mathbf{a} - \mathbf{a}^i\|_\infty \leq \epsilon \text{ for all } i \geq N.$$

This also entails that $\mathbf{a}$ is bounded. So l^∞ is complete.

To prove that c_0 is complete we need to show that a uniform limit of a sequence of elements of c_0 is again in c_0. So suppose $\mathbf{a}^i \to \mathbf{a}$ in sup norm and each $\mathbf{a}^i$ is in c_0. Then for any $\epsilon > 0$ we can find i such that $\|\mathbf{a} - \mathbf{a}^i\|_\infty \le \epsilon$, and $|a_n^i|$ is eventually $\le \epsilon$ so $|a_n|$ must be eventually $\le 2\epsilon$. As ϵ was arbitrary, this shows that $\mathbf{a} \in c_0$. $\qquad\square$

The spaces l^1, l^2, l^∞, and c_0 are our basic repertoire of infinite-dimensional Banach spaces. They can also be denoted $l^1(\mathbf{N})$, $l^2(\mathbf{N})$, etc. to distinguish them from similar spaces like $l^1(\mathbf{Z})$ or $l^\infty(\mathbf{N}^2)$ whose elements are indexed by other sets than $\mathbf{N}$. Also, for $n \in \mathbf{N}$ we can identify $\mathbf{F}^n$ with the set of scalar sequences which satisfy $a_i = 0$ for $i > n$ and let it inherit a norm from l^1, l^2, or l^∞. We denote the resulting spaces by l_n^1, l_n^2, and l_n^∞. The unit balls of l_2^1, l_2^2, and l_2^∞ in the real case are pictured in Figure 3.2.

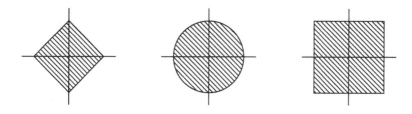

Fig. 3.2   Unit balls of real l_2^1, l_2^2, and l_2^∞

As we emphasized in Chapter 1, we generally prefer to work with separable metric spaces. So we immediately want to determine which of the above spaces are separable. The spaces l_n^1, l_n^2, and l_n^∞ are clearly separable because $\mathbf{F}_0^n$ is dense in each of them. In the infinite-dimensional case we have the following result.

Proposition 3.9. *The spaces c_0, l^1, and l^2 are separable. The space l^∞ is nonseparable.*

Proof.   If $\mathbf{b}$ belongs to c_0, l^1, or l^2 then the norm of its "tail" $\mathbf{b}'$ with entries

$$b_n' = \begin{cases} 0 & \text{if } n \le N \\ b_n & \text{if } n > N \end{cases}$$

goes to zero as $N \to \infty$. It follows that every element of the space is approximated by elements which are constantly zero past some point. Thus

$\bigcup_n l_n^\infty$, $\bigcup_n l_n^1$, and $\bigcup_n l_n^2$ are respectively dense in c_0, l^1, and l^2. Since a countable union of countable sets is countable, separability of the latter spaces now follows from separability of l_n^∞, l_n^1, and l_n^2 for each n.

To see that l^∞ is nonseparable, consider the sequences all of whose entries are either 0 or 1. There are uncountably many such sequences, and the sup norm distance between any two of them is 1. Thus the balls of radius $\frac{1}{2}$ about these sequences constitute an uncountable family of disjoint open balls in l^∞, and this shows that l^∞ cannot be separable. □

Although l^∞ is not separable as a Banach space, we will see in Chapter 4 that it enjoys a different kind of separability property.

Completeness and separability will always be among our first concerns when we introduce a normed vector space. We close this section with two general tools for verifying these properties. Say that a series $\sum x_n$ of elements of a normed vector space E is *summable* if the partial sums $\sum_{k=1}^n x_k$ converge to a limit x in E as $n \to \infty$ (in which case we write $x = \sum_{n=1}^\infty x_n$), and *absolutely summable* if the sum $\sum_{n=1}^\infty \|x_n\|$ is finite.

Lemma 3.10. *Let E be a normed vector space. Then E is a Banach space iff every absolutely summable series in E is summable.*

Proof. ($\Rightarrow$) Suppose E is complete and let $\sum x_n$ be an absolutely summable series. For each $n \in \mathbf{N}$ let s_n be the partial sum $s_n = \sum_{k=1}^n x_k$. Given $\epsilon > 0$, find $N \in \mathbf{N}$ such that $\sum_{k=N+1}^\infty \|x_k\| \le \epsilon$; then for any $m > n \ge N$ we have

$$\|s_m - s_n\| = \|x_{n+1} + x_{n+2} + \cdots + x_m\| \le \sum_{k=N+1}^\infty \|x_k\| \le \epsilon.$$

Thus the sequence (s_n) is Cauchy. Since E is complete, (s_n) converges, i.e., $\sum x_n$ is summable.

($\Leftarrow$) Suppose every absolutely summable series is summable and let (x_n) be a Cauchy sequence in E. It is enough to show that some subsequence of (x_n) converges. Thus, by passing to a subsequence we may assume $\|x_{n+1} - x_n\| \le 2^{-n}$ for all n. Define a new sequence (y_n) by setting $y_1 = x_1$ and $y_{n+1} = x_{n+1} - x_n$ for $n \ge 2$. Then $\sum y_n$ is absolutely summable since

$$\sum_{n=1}^\infty \|y_n\| \le \|y_1\| + \sum_{n=1}^\infty 2^{-n} < \infty,$$

and by hypothesis it follows that $\sum y_n$ is summable. But (x_n) is the sequence of partial sums of this series, so this means that (x_n) converges. □

The pedestrian way to see that we can extract a rapidly converging subsequence from (x_n) in the preceding proof is to find n_1 such that $m, n \geq n_1$ implies $\|x_m - x_n\| \leq \frac{1}{2}$, then find $n_2 > n_1$ such that $m, n \geq n_2$ implies $\|x_m - x_n\| \leq \frac{1}{4}$, etc. A slightly slicker method is to pass to the metric completion of E in which we can be sure the Cauchy sequence (x_n) has a limit x, then find an increasing sequence (n_k) such that $d(x, x_{n_k}) \leq 2^{-k-1}$ for all k.

Lemma 3.10 is useful because it allows us to take advantage of the linear structure of E when verifying completeness. A similar comment can be made about the next result.

Lemma 3.11. *A normed vector space is separable iff it contains a sequence of vectors whose linear span is dense.*

Proof. Let E be a normed vector space. If E is separable then there is a sequence of vectors which is dense in E, so certainly its span is dense. Conversely, suppose there is a sequence whose span is dense in E. Then the finite linear combinations of the vectors in the sequence with coefficients in $\mathbf{F}_0$ constitute a countable dense subset of E, so E is separable. □

3.2 Basic constructions

Now that we have defined Banach spaces, the next order of business is to look at functions between them. Linear maps are the natural functions between vector spaces, but when the vector spaces involved are equipped with norms, we typically want to impose a continuity requirement too. Our first result states that continuity is equivalent to a boundedness condition.

When $T : E \to F$ is a linear map and $x \in E$, we usually write Tx instead of $T(x)$. This standard notation probably comes from identifying T with a matrix, x with a column vector, and $T(x)$ with the matrix product of T and x.

Proposition 3.12. *Let E and F be normed vector spaces and let $T : E \to F$ be a linear map. Then T is continuous iff there exists a constant $C \leq 0$ such that $\|Tx\| \leq C\|x\|$ for all $x \in E$.*

Proof. ($\Rightarrow$) Suppose T is continuous for the norm topologies on E and F. The open unit ball $(F)_1$ is an open set in F, so $T^{-1}((F)_1)$ must be an open set in E. Since $T(0) = 0$, this set contains the origin of E, and hence it must contain $(E)_r$ for some $r > 0$.

Choose $C > 0$ such that $\frac{1}{C} < r$. Then for any nonzero $x \in E$, the vector $y = \frac{1}{C\|x\|}x$ satisfies $\|y\| < r$, i.e., $y \in (E)_r$. Hence $Ty \in (F)_1$, and we conclude that

$$\|Tx\| = C\|x\| \cdot \|Ty\| < C\|x\|.$$

($\Leftarrow$) Now suppose there exists $C > 0$ such that $\|Tx\| \leq C\|x\|$ for all $x \in E$. Fix $x \in E$ and $\epsilon > 0$; we must find $\delta > 0$ such that

$$\|x - y\| \leq \delta \quad \Rightarrow \quad \|Tx - Ty\| \leq \epsilon.$$

Taking $\delta = \frac{\epsilon}{C}$ suffices because then

$$\|x - y\| \leq \delta \quad \Rightarrow \quad \|Tx - Ty\| = \|T(x - y)\| \leq C\|x - y\| \leq \epsilon.$$

So T is continuous. $\qquad\qquad\qquad\qquad\qquad\qquad\qquad\qquad\qquad\qquad\square$

This motivates the following definition.

Definition 3.13. A linear map $T : E \to F$ between normed vector spaces is *bounded* if there exists $C \geq 0$ such that $\|Tx\| \leq C\|x\|$ for all $x \in E$. It is an *isomorphism* if it is a bijection and both T and T^{-1} are bounded. It is an *isometric isomorphism* if it is bijective and isometric.

Just as with metric spaces, there are two natural notions of "being the same" for normed vector spaces: in this case, linear homeomorphism and linear isometry. Since we are more interested in linear isometries, we will use the notation $E \cong F$ to mean that E and F are isometrically isomorphic.

We say that two norms on the same vector space are *equivalent* if their norm topologies are the same. This can also be expressed by saying that the identity map is an isomorphism from the space equipped with one norm to the space equipped with the other norm.

We need a notation for the minimal constant that verifies the boundedness condition for a bounded linear map.

Definition 3.14. The *operator norm* of a bounded linear map $T : E \to F$ between normed vector spaces is the quantity

$$\|T\| = \inf\{C \geq 0 : \|Tx\| \leq C\|x\| \text{ for all } x \in E\}$$
$$= \sup\{\|Tx\| : \|x\| \leq 1\}.$$

As this terminology and notation suggests, the operator norm is indeed a norm on the space of bounded linear maps from E to F. We will return to this point later in this section.

It is an exercise to verify that the two formulas for $\|T\|$ agree. We also leave the proof of the next result as an exercise.

Proposition 3.15. *Let $T_1 : E_1 \to E_2$ and $T_2 : E_2 \to E_3$ be bounded linear maps between normed vector spaces. Then $\|T_2 \circ T_1\| \leq \|T_1\| \|T_2\|$.*

A key point is that any continuous linear map between normed vector spaces is automatically uniformly continuous: for any $x \in E$ and any $\epsilon > 0$ we have $T(\text{ball}_\epsilon(x)) \subseteq \text{ball}_{\epsilon\|T\|}(Tx)$. It follows from uniform continuity that a bounded linear map from a dense subspace of E into a Banach space can always be extended to all of E. (We emphasize that by "subspace" we will always mean "linear subspace".) This is basically a metric space fact but we include a proof anyway.

Proposition 3.16. *Let E_0 be a dense subspace of a normed vector space E and let F be a Banach space. Then any bounded linear map from E_0 to F extends uniquely to a bounded linear map from E to F. The operator norm does not increase.*

Proof. Let $T : E_0 \to F$ be bounded. For each $x \in E$ find a sequence (x_n) in E_0 that converges to x and define $\tilde{T}x = \lim Tx_n$. Since (x_n) is Cauchy and T is bounded, the sequence (Tx_n) is also Cauchy and hence the limit makes sense. If (x'_n) is another sequence that converges to x then $\|x_n - x'_n\| \to 0$, and hence $\|Tx_n - Tx'_n\| = \|T(x_n - x'_n)\| \to 0$, so the value is well-defined. Since continuity requires that $\tilde{T}x_n \to \tilde{T}x$, it is clear that the extension is unique.

If $x_n \to x$ and $a \in \mathbf{F}$ then $ax_n \to ax$, so that
$$\tilde{T}(ax) = \lim T(ax_n) = \lim(a \cdot Tx_n) = a\tilde{T}x;$$
if $y_n \to y$ as well then $x_n + y_n \to x + y$ (triangle inequality) so that
$$\tilde{T}(x + y) = \lim T(x_n + y_n) = \lim(Tx_n + Ty_n) = \tilde{T}x + \tilde{T}y.$$
So the extension is linear. It is bounded with $\|\tilde{T}\| \leq \|T\|$ because
$$\|\tilde{T}x\| = \lim \|Tx_n\| \leq \|T\| \cdot \lim \|x_n\| = \|T\| \|x\|$$
for all $x \in E$. $\qquad\square$

Note that since $\tilde{T}$ extends T, the inequality $\|T\| \leq \|\tilde{T}\|$ is automatic. So we have $\|T\| = \|\tilde{T}\|$.

We will now describe some elementary ways of constructing new Banach spaces from old ones. The simplest method is passing to a subspace.

Proposition 3.17. *Let E be a Banach space and let E_0 be a closed subspace of E. Then E_0 is a Banach space with the inherited norm.*

This is essentially just a special case of the easy fact that a closed subset of a complete metric space is always complete.

As simple as it is, this closed subspace construction provides us with many new examples of Banach spaces. Indeed, *every* separable Banach space appears as a closed subspace of l^∞. (This will be an exercise; it uses the Hahn-Banach theorem.) At the other extreme, l^2 contains very few Banach spaces; all infinite-dimensional closed subspaces of l^2 are isometrically isomorphic to each other, as are any two subspaces of the same finite dimension. We will prove this in Corollary 5.9.

We can also take the quotient by a closed subspace, a slightly less trivial construction. We use the notation $x + E_0 = \{x + y : y \in E_0\}$ for a coset of a subspace E_0 of a vector space E, i.e., an element of the quotient space E/E_0.

Definition 3.18. Let E be a normed vector space and let E_0 be a closed subspace of E. We define the *quotient norm* on the quotient space E/E_0 by

$$\|x + E_0\| = \inf\{\|z\| : z \in x + E_0\} = \inf\{\|x + y\| : y \in E_0\}.$$

This formula can be understood geometrically as the distance from the origin of E to the affine subspace $x + E_0$. Notice that since E_0 is a subspace we could equally well use the formula $\inf\{\|x - y\| : y \in E_0\}$. Thus $\|x + E_0\|$ can also be pictured as the distance from x to E_0.

Proposition 3.19. *Let E_0 be a closed subspace of a normed vector space E. Then the quotient norm is a norm on E/E_0. If E_0 is a proper subspace then the quotient map $\pi : E \to E/E_0$ has norm 1. If E is a Banach space then so is E/E_0.*

Proof. The first two properties of a norm are easy to check in E/E_0. To verify the triangle inequality, let $x, x' \in E$ and observe that for any $y, y' \in E_0$ we have

$$\|(x + x') + (y + y')\| \le \|x + y\| + \|x' + y'\|.$$

As every expression on the left contributes to the infimum which defines the quotient norm of $(x + x') + E_0$, taking the infimum over y and y' yields $\|(x + x') + E_0\| \le \|x + E_0\| + \|x' + E_0\|$, as desired.

It is clear that $\pi : x \mapsto x + E_0$ has norm at most 1, since $\|x + E_0\| \le \|x + 0\| = \|x\|$. Conversely, if E_0 is a proper subspace, i.e., $E_0 \ne E$, then fix any $x \in E \setminus E_0$. For any $\epsilon > 0$, find $y \in E_0$ such that $\|x + y\| \le (1 + \epsilon)\|x + E_0\|$;

then $\|\pi(x+y)\| = \|x+E_0\| \neq 0$ implies that $\|\pi\| \geq \frac{1}{1+\epsilon}$. As ϵ was arbitrary, it follows that $\|\pi\| = 1$.

We use Lemma 3.10 to verify completeness. Thus, suppose E is a Banach space and let $\sum(x_n + E_0)$ be an absolutely summable series in E/E_0. For each n, choose $y_n \in E_0$ such that $\|x_n + y_n\| \leq \|x_n + E_0\| + 2^{-n}$. Then the series $\sum(x_n + y_n)$ is absolutely summable in E (since $\sum\|x_n + E_0\|$ and $\sum 2^{-n}$ are both summable), and hence it converges to some value x. We then have

$$\left\|(x + E_0) - \sum_{k=1}^{n}(x_k + E_0)\right\| \leq \left\|x - \sum_{k=1}^{n}(x_k + y_k)\right\| \to 0$$

as $n \to \infty$. Thus $\sum(x_n + E_0)$ sums to $x + E_0$ in E/E_0. Lemma 3.10 now implies that E/E_0 is complete. $\qquad\square$

We mentioned earlier that every separable Banach space embeds in l^∞. Every separable Banach space can also be realized as a quotient of l^1. This will be an exercise too.

Quotient spaces enjoy a simple universal property.

Proposition 3.20. *Let $T : E \to F$ be a bounded linear map between normed vector spaces. Then there is a bounded linear map $\tilde{T} : E/\ker T \to F$ with $\|\tilde{T}\| = \|T\|$ such that $T = \tilde{T} \circ \pi$, where $\pi : E \to E/\ker T$ is the quotient map.*

Proof. Define $\tilde{T}(x + \ker T) = T(x)$ for any $x \in E$. It is easy to see that $\tilde{T}$ is well-defined, that $T = \tilde{T} \circ \pi$, and that $\tilde{T}$ is linear. Next, let $x \in E$. For any $\epsilon > 0$ find $y \in \ker T$ such that $\|x + y\| \leq (1 + \epsilon)\|x + \ker T\|$; then

$$\|\tilde{T}(x + \ker T)\| = \|Tx\| = \|T(x + y)\| \leq \|T\| \cdot (1 + \epsilon)\|x + \ker T\|.$$

Since ϵ and x were arbitrary, we conclude that $\|\tilde{T}\| \leq \|T\|$. Conversely, $T = \tilde{T} \circ \pi$ implies that $\|T\| \leq \|\tilde{T}\|$ by Proposition 3.15 and Proposition 3.19. So $\|T\| = \|\tilde{T}\|$. $\qquad\square$

Next we consider direct sums. There are a variety of versions of this construction. The most important are listed in the following definition.

Definition 3.21. Let (E_n) be a finite or infinite sequence of normed vector spaces.

(a) Its l^1 *direct sum* is the space

$$\bigoplus{}^1 E_n = \left\{\mathbf{x} \in \prod E_n : \sum\|x_n\| < \infty\right\},$$

equipped with the norm $\|\mathbf{x}\|_1 = \sum\|x_n\|$.

(b) Its l^2 *direct sum* is the space

$$\bigoplus\nolimits^2 E_n = \left\{ \mathbf{x} \in \prod E_n : \sum \|x_n\|^2 < \infty \right\},$$

equipped with the norm $\|\mathbf{x}\|_2 = (\sum \|x_n\|^2)^{1/2}$.

(c) Its l^∞ *direct sum* is the space

$$\bigoplus\nolimits^\infty E_n = \left\{ \mathbf{x} \in \prod E_n : \sup \|x_n\| < \infty \right\},$$

equipped with the norm $\|\mathbf{x}\|_\infty = \sup \|x_n\|$.

(d) If the sequence (E_n) is infinite then its c_0 *direct sum* is the subspace

$$\bigoplus\nolimits_0 E_n = \left\{ \mathbf{x} \in \prod E_n : \|x_n\| \to 0 \right\}$$

of $\bigoplus^\infty E_n$, with the inherited norm.

Note that if (E_n) is a finite sequence, then the underlying space of the l^1, l^2, and l^∞ constructions is the full product $\prod E_n$. However, the norms are different.

Also, it should be clear that the spaces l^1, l^2, l^∞, and c_0 of Examples 3.3, 3.5, and 3.7 are the special cases of Definition 3.21 that arise when we take the direct sum of a sequence of copies of $\mathbf{F}$.

Proposition 3.22. *Let (E_n) be a finite or infinite sequence of normed vector spaces. Then the direct sums in Definition 3.21 are also normed vector spaces. If the E_n are Banach spaces then so are each of their direct sums. If each E_n is separable then so are their l^1, l^2, and c_0 direct sums; the l^∞ direct sum of any sequence of nonzero normed vector spaces is nonseparable.*

We omit the proof of this result. It follows the proofs of Propositions 3.4, 3.6, 3.8, and 3.9 almost identically, except for the proof that $\|\cdot\|_2$ is a norm on the l^2 direct sum, which we leave as an exercise.

There is one more construction of basic importance. The set of bounded linear maps from one normed vector space to another is itself a normed vector space, with operations defined pointwise. This construction is of interest primarily in two special cases, when the target space is the scalar field (which we will discuss in the next section and say more about in Chapter 4) and when the range equals the domain and is a Hilbert space (which we will discuss in Chapter 5).

We denote by $\mathcal{B}(E, F)$ the set of bounded linear maps from E into F.

Proposition 3.23. *Let E and F be normed vector spaces. Then $\mathcal{B}(E, F)$ is a vector space with operations defined pointwise, and the operator norm is a norm on $\mathcal{B}(E, F)$. If F is a Banach space then so is $\mathcal{B}(E, F)$.*

Proof. The pointwise operations are given by $(S+T)(x) = Sx + Tx$ and $(aT)(x) = a \cdot Tx$ for $x \in E$ and $a \in \mathbf{F}$. It is clear that a scalar multiple of a bounded linear map is bounded, and the sum of two bounded linear maps is bounded because for any $x \in E$ we have

$$\|(S+T)(x)\| = \|Sx + Tx\| \leq \|Sx\| + \|Tx\| \leq (\|S\| + \|T\|)\|x\|;$$

this implies that $S+T$ is bounded with $\|S+T\| \leq \|S\| + \|T\|$. The operator norm trivially satisfies the other properties of a norm.

Now suppose F is a Banach space and let (T_n) be a Cauchy sequence of bounded linear maps. For each $x \in E$ we have

$$\|T_m x - T_n x\| = \|(T_m - T_n)(x)\| \leq \|T_m - T_n\|\|x\|,$$

which shows that the sequence $(T_n x)$ is Cauchy in F. Define $T : E \to F$ by setting $Tx = \lim T_n x$. Then T is linear because

$$T(ax) = \lim_{n \to \infty} T_n(ax) = \lim_{n \to \infty} (a \cdot T_n x) = a \cdot Tx$$

and

$$T(x + y) = \lim_{n \to \infty} T_n(x + y) = \lim_{n \to \infty} (T_n x + T_n y) = Tx + Ty.$$

It is bounded because $\|Tx\| = \lim \|T_n x\| \leq \sup_n \|T_n\|\|x\|$ (and $\sup \|T_n\|$ is finite since (T_n) is Cauchy). Finally, to verify $T_n \to T$, let $\epsilon > 0$ and find $N \in \mathbf{N}$ such that $\|T_m - T_n\| \leq \epsilon$ for all $m, n \geq N$. Then for any $n \geq N$ and any $x \in E$ we have

$$\|(T - T_n)x\| = \lim_{m \to \infty} \|(T_m - T_n)x\| \leq \epsilon\|x\|,$$

i.e., $\|T - T_n\| \leq \epsilon$. This shows that T_n converges to T. $\square$

3.3 The Hahn-Banach theorem

The Hahn-Banach theorem deals with bounded linear maps into $\mathbf{F}$. We have a special name for such maps.

Definition 3.24. A linear map from a vector space into $\mathbf{F}$ is called a *linear functional*. The *dual space* of a normed vector space E is the Banach space $E' = \mathcal{B}(E, \mathbf{F})$ of all bounded linear functionals on E.

Dual spaces of nonseparable Banach spaces tend to be quite pathological. However, dual spaces of separable Banach spaces are extremely useful, and one can often describe them in very concrete terms. Here are some simple examples.

For $n \in \mathbf{N}$ let $\mathbf{e}^n$ be the sequence whose nth entry is 1 and whose other entries are all 0.

Theorem 3.25. *We have*

   (i) $(c_0)' \cong l^1$;
   (ii) $(l^1)' \cong l^\infty$;
   (iii) $(l^2)' \cong l^2$.

Proof.   (i) For each $\mathbf{b} \in l^1$ define $T\mathbf{b} \in (c_0)'$ by setting $(T\mathbf{b})(\mathbf{a}) = \sum a_n b_n$ for all $\mathbf{a} \in c_0$. Observe that this series is absolutely summable because

$$\sum |a_n b_n| \le \|\mathbf{a}\|_\infty \sum |b_n| = \|\mathbf{a}\|_\infty \|\mathbf{b}\|_1.$$

Also, it is straightforward to check that $T\mathbf{b}$ is a linear map from c_0 to $\mathbf{F}$. So $T\mathbf{b}$ is indeed a bounded linear functional on c_0 which, by the preceding inequality, satisfies $\|T\mathbf{b}\| \le \|\mathbf{b}\|_1$. For the converse inequality, let $m \in \mathbf{N}$ and consider the element $\mathbf{a}^m$ of c_0 defined by

$$\mathbf{a}^m = \left(\frac{|b_1|}{b_1}, \ldots, \frac{|b_m|}{b_m}, 0, 0, \ldots \right) = \frac{|b_1|}{b_1}\mathbf{e}^1 + \cdots + \frac{|b_m|}{b_m}\mathbf{e}^m,$$

using the convention that $\frac{0}{0} = 0$. Then

$$(T\mathbf{b})(\mathbf{a}^m) = \sum_{n=1}^m |b_n|,$$

and taking $m \to \infty$ shows that $\|T\mathbf{b}\| \ge \|\mathbf{b}\|_1$. As linearity of T is straightforward, we now have that T is an isometric isomorphism from l^1 into $(c_0)'$. To prove surjectivity, let $f \in (c_0)'$ be arbitrary. For each n define $b_n = f(\mathbf{e}^n)$. Then the resulting sequence $\mathbf{b}$ belongs to l^1 because, defining $\mathbf{a}^m$ as above, we have

$$\sum_{n=1}^m |b_n| = |f(\mathbf{a}^m)| \le \|f\|$$

for all m. Now $T\mathbf{b}$ is a bounded linear functional which satisfies $T\mathbf{b}(\mathbf{e}^n) = f(\mathbf{e}^n)$ for all n; since the span of the $\mathbf{e}^n$ is dense in c_0, this shows that $T\mathbf{b} = f$. So T is surjective.

   (ii) Exercise.

   (iii) Recall the inner product we defined on l^2 just after Proposition 3.6. For each $\mathbf{a}$ and $\mathbf{b}$ in l^2 set $(T\mathbf{b})(\mathbf{a}) = \sum a_n b_n = \langle \mathbf{a}, \bar{\mathbf{b}} \rangle$, where $\bar{\mathbf{b}}$ is the pointwise complex conjugate sequence with entries $\bar{b}_n$. This is clearly a linear functional, and it is bounded with $\|T\mathbf{b}\| \le \|\bar{\mathbf{b}}\|_2 = \|\mathbf{b}\|_2$ by the inequality $|\langle \mathbf{a}, \bar{\mathbf{b}} \rangle| \le \|\mathbf{a}\|_2 \|\bar{\mathbf{b}}\|_2$ which we proved when we introduced the inner product. We can achieve the reverse inequality by taking $\mathbf{a} = \bar{\mathbf{b}}$: we have $(T\mathbf{b})(\bar{\mathbf{b}}) = \|\mathbf{b}\|_2^2$, which implies that $\|T\mathbf{b}\| \geq \|\mathbf{b}\|_2$. Linearity of the map T is straightforward, so it is an isometric isomorphism from l^2 into $(l^2)'$. To prove surjectivity, let $f \in (l^2)'$ and define $b_n = f(\mathbf{e}^n)$. Fixing $m \in \mathbf{N}$, set

$$\mathbf{a}^m = (\bar{b}_1, \ldots, \bar{b}_m, 0, 0, \ldots) = \bar{b}_1 \mathbf{e}^1 + \cdots + \bar{b}_m \mathbf{e}^m;$$

then $\|\mathbf{a}^m\|_2 = (\sum_{n=1}^m |b_n|^2)^{1/2}$ and $|f(\mathbf{a}^m)| = \|\mathbf{a}^m\|_2^2$, which implies that $\|\mathbf{a}^m\|_2 \leq \|f\|$. Taking $m \to \infty$ then shows that $\mathbf{b} \in l^2$, and since $T\mathbf{b}$ agrees with f on the $\mathbf{e}^n$, whose span is dense in l^2, it follows that $T\mathbf{b} = f$. □

We now turn to the Hahn-Banach theorem, the single most important result in the theory of Banach spaces. We saw in Proposition 3.16 that a bounded linear map defined on a dense subspace can be continuously extended to the whole space without increasing its norm. The Hahn-Banach theorem says that if the target space is $\mathbf{F}$ then a bounded linear map defined on *any* subspace can be extended to the whole space without increasing its norm.

The significance of this fact is that it gives us the ability to analyze a Banach space using bounded linear functionals. For instance, it follows from the Hahn-Banach theorem that if $x, y \in E$ are distinct then there exists $f \in E'$ such that $f(x) \neq f(y)$ (apply Corollary 3.28 to $x - y$). Or if E_0 is a closed subspace and $x \notin E_0$ then there exists $f \in E'$ satisfying $f|_{E_0} = 0$ and $f(x) \neq 0$ (see Corollary 3.29).

Our proof of the Hahn-Banach theorem has a somewhat geometric flavor, and it requires the use of real scalars. The following lemma shows why this is sufficient. In this result we make use of the fact that any complex vector space E can also be considered to be a real vector space. Define the real and imaginary parts of a complex linear map $f : E \to \mathbf{F}$ by $(\Re f)(x) = \Re f(x)$ and $(\Im f)(x) = \Im f(x)$.

Lemma 3.26. *Let E be a complex normed vector space. Then the real part of any bounded complex linear functional on E is a bounded real linear functional with the same norm, and every bounded real linear functional is the real part of a unique bounded complex linear functional. This establishes an isometry between the dual of E as a real normed vector space and its dual as a complex normed vector space.*

Proof. Let $f : E \to \mathbf{C}$ be a bounded complex linear functional. It is easy to see that $\Re f$ is a real linear functional and that $\|\Re f\| \leq \|f\|$. For the reverse inequality, let $x \in E$ with $f(x) \neq 0$ and define $a = \frac{|f(x)|}{f(x)}$. Then $\|ax\| = \|x\|$ and $(\Re f)(ax) = |f(x)|$. This shows that the norm of $\Re f$ is no smaller than the norm of f.

Now let $g : E \to \mathbf{R}$ be a bounded real linear functional. Define $f : E \to \mathbf{C}$ by $f(x) = g(x) - ig(ix)$. It is straightforward to verify that f is complex linear. (First check that $f(x + y) = f(x) + f(y)$, then that $f(ax) = af(x)$ for $a \in \mathbf{R}$, and then that $f(ix) = if(x)$.) Thus, any bounded real linear functional on E is the real part of a bounded complex linear functional. The latter is unique because $\Im f(x) = \Re f(ix)$, so if the real parts of two linear functionals agree then their imaginary parts must also agree.

The last statement follows immediately. $\square$

We can now prove the Hahn-Banach theorem. For the sake of simplicity we restrict to the separable case.

Theorem 3.27. *(Hahn-Banach theorem) Let E be a separable normed vector space, E_0 a subspace of E, and $f : E_0 \to \mathbf{F}$ a bounded linear functional. Then f extends to a bounded linear functional $\tilde{f} : E \to \mathbf{F}$ such that $\|\tilde{f}\| = \|f\|$.*

Proof.   First, note that $\|\tilde{f}\| \geq \|f\|$ is automatic given that $\tilde{f}$ extends f; the point is to avoid increasing the norm. Assume real scalars. The key step is to show that we can extend a bounded linear functional to one additional dimension without increasing its norm. Thus let $x \in E \setminus E_0$. By linearity, to extend f to $\mathrm{span}(E_0 \cup \{x\})$ we only have to specify the value of $\tilde{f}(x)$. Since every element of $\mathrm{span}(E_0 \cup \{x\})$ is of the form $ax + y$ for some $a \in \mathbf{R}$ and some $y \in E_0$, the norm of $\tilde{f}$ will not increase provided

$$|\tilde{f}(ax + y)| \leq \|f\| \|ax + y\|$$

for all $a \in \mathbf{R}$ and $y \in E_0$. Thus, the value $t = \tilde{f}(x)$ to be chosen must satisfy

$$-\|f\| \|ax + y\| \leq at + f(y) \leq \|f\| \|ax + y\|.$$

For $a = 0$ this condition is automatic, and for nonzero a we can set $z = \frac{1}{a}y$ and isolate t, which yields

$$-\|f\| \|x + z\| - f(z) \leq t \leq \|f\| \|x + z\| - f(z). \qquad (*)$$

We need these inequalities to hold for all $z \in E_0$. (Note that we get the same inequalities regardless of whether a is positive or negative.) Now observe that for any $z_1, z_2 \in E_0$ we have

$$f(z_2 - z_1) \leq \|f\| \|z_2 - z_1\| \leq \|f\| \|x + z_2\| + \|f\| \|x + z_1\|.$$

Rearranging yields

$$-\|f\|\|x + z_1\| - f(z_1) \le \|f\|\|x + z_2\| - f(z_2),$$

and as this holds for all z_1 and z_2 in E_0 we conclude that

$$\sup_{z \in E_0} \left(-\|f\|\|x + z\| - f(z)\right) \le \inf_{z \in E_0} \left(\|f\|\|x + z\| - f(z)\right).$$

The inequalities in $(*)$ will then be satisfied by any value of t that lies between these bounds. This shows that a suitable value of t is available. Thus, we are able to extend f to one additional dimension without increasing its norm.

Now let (x_n) be a sequence in E whose span is dense, and for each $n \in \mathbf{N}$ let $E_n = \mathrm{span}(E_0 \cup \{x_1, \ldots, x_n\})$. By the preceding argument we can successively extend from E_n to E_{n+1} without increasing norm for all n, and this ultimately yields a norm preserving extension of f to $\bigcup E_n$, which is dense in E. The proof of the real case is completed by appealing to Proposition 3.16.

In the case of complex scalars, regard E as a real normed vector space and let $g = \Re f$. The first part of the proof implies that there is a norm preserving real extension $\tilde{g}$ of g to E. We can then apply Lemma 3.26 to find a complex linear functional $\tilde{f}$ on E whose real part is $\tilde{g}$ and whose norm equals $\|f\|$. By the uniqueness part of Lemma 3.26, f and $\tilde{f}$ must agree on E_0. $\qquad\square$

We collect some easy consequences of the Hahn-Banach theorem.

Corollary 3.28. *Let E be a normed vector space and let $x \in E$ be nonzero. Then there exists $f \in E'$ such that $\|f\| = 1$ and $f(x) = \|x\|$.*

Proof. Let $E_0 = \mathbf{F}x$ be the span of x and define $f(ax) = a\|x\|$ for all $a \in \mathbf{F}$. This linear functional on E_0 clearly has norm 1, so it extends to a norm 1 linear functional on E by the Hahn-Banach theorem. $\qquad\square$

Recall the quotient norm introduced in Definition 3.18.

Corollary 3.29. *Let E be a normed vector space, $E_0 \subset E$ a proper closed subspace, and $x \in E \setminus E_0$. Then there exists $f \in E'$ such that $\|f\| = 1$, $f|_{E_0} = 0$, and $f(x) = \|x + E_0\|$.*

Proof. Define f on $\mathrm{span}(E_0 \cup \{x\})$ by setting $f(ax + y) = a\|x + E_0\|$ for all $a \in \mathbf{F}$ and $y \in E_0$. Then $f|_{E_0} = 0$ and $f(x) = \|x + E_0\|$ are immediate. Also, we have $|f(ax+y)| = \|ax + E_0\| \le \|ax + y\|$ for all a and y, so $\|f\| \le 1$.

Conversely, by the definition of the quotient norm there is a sequence (y_n) in E_0 such that $\|x + E_0\| = \lim \|x + y_n\|$, so

$$\frac{f(x + y_n)}{\|x + y_n\|} = \frac{\|x + E_0\|}{\|x + y_n\|} \to 1;$$

this shows that $\|f\| \geq 1$, and we conclude that $\|f\| = 1$. We complete the proof by using the Hahn-Banach theorem to extend f to all of E. $\square$

In the preceding result, recall that $\|x + E_0\|$ is the distance from x to E_0, so if $\|f\| = 1$ and $f|_{E_0} = 0$ then the value of $f(x)$ cannot be more than $\|x + E_0\|$. So the conclusion is optimal.

We need to establish some notation for the next corollary.

Definition 3.30. For any element x of a normed vector space E let $\hat{x}$: $E' \to \mathbf{F}$ be the linear functional defined by

$$\hat{x}(f) = f(x).$$

We can call $\hat{x}$ "evaluation at x". It is linear just by the way the operations in E' are defined, e.g., $(f + g)(x) = f(x) + g(x)$.

Corollary 3.31. *Let E be a normed vector space. Then the map $x \to \hat{x}$ is a linear isometry from E into E''.*

Proof. For any x the linear functional $\hat{x}$ is bounded with $\|\hat{x}\| \leq \|x\|$ because

$$|\hat{x}(f)| = |f(x)| \leq \|f\|\|x\|$$

for all $f \in E'$. Thus the map $x \mapsto \hat{x}$ takes E into E''. Linearity of this map follows from the linearity of elements of E', e.g.,

$$(\widehat{x + y})(f) = f(x + y) = f(x) + f(y) = \hat{x}(f) + \hat{y}(f).$$

Finally, for any nonzero $x \in E$ we can find $f \in E'$ as in Corollary 3.28. Then since $\|f\| = 1$ and

$$|\hat{x}(f)| = |f(x)| = \|x\|,$$

it follows that $\|\hat{x}\| \geq \|x\|$. We conclude that $\|\hat{x}\| = \|x\|$ for all $x \in E$. $\square$

A Banach space is called *reflexive* if the map of Corollary 3.31 is surjective. For instance, every finite-dimensional space is reflexive. We can also see from Theorem 3.25 that c_0 is not reflexive but l^2 is. Note that an incomplete normed vector space can never be reflexive, since according to Proposition 3.23 dual spaces are always complete. This suggests a way to tie up a loose end from Section 3.1: for any normed vector space E the space E'' is complete, so it follows from Corollary 3.31 that E isometrically embeds in a Banach space.

Corollary 3.32. *Every normed vector space can be isometrically and densely embedded in a Banach space.*

To achieve dense embedding, just take the target space to be the closure of the image of E in E''. This shows that any normed vector space can be completed to a Banach space.

3.4 The Banach isomorphism theorem

In this section we present another standard tool, the Banach isomorphism theorem, and a medley of consequences: the open mapping and closed graph theorems, the principle of uniform boundedness, and the Banach-Steinhaus theorem.

We require two ingredients, the Baire category theorem and the approximation lemma. Both of these results are useful in other settings as well.

Theorem 3.33. *(Baire category theorem) Let X be a complete metric space and let (U_n) be a sequence of dense open subsets of X. Then $\bigcap U_n$ is dense in X.*

Proof. Let V be an open set; we must show that $V \cap \bigcap U_n$ is not empty. Since U_1 is dense, $V \cap U_1$ is a nonempty open set, so it contains a ball, say $\text{ball}_{r_1}(x_1)$. Possibly decreasing the value of r_1, we may assume that $r_1 \leq 1$ and that the closure of $\text{ball}_{r_1}(x_1)$ is contained in $V \cap U_1$. Since U_2 is dense, $\text{ball}_{r_1}(x_1) \cap U_2$ is now a nonempty open set, so by a similar argument we can find $r_2 \leq \frac{1}{2}$ and a ball, say $\text{ball}_{r_2}(x_2)$, whose closure is contained in $\text{ball}_{r_1}(x_1) \cap U_2$. Proceed inductively to obtain a sequence $(\text{ball}_{r_n}(x_n))$ such that

$$r_n \leq \frac{1}{n} \quad \text{and} \quad \overline{\text{ball}_{r_n}(x_n)} \subseteq \text{ball}_{r_{n-1}}(x_{n-1}) \cap U_n$$

for all $n \geq 2$. Then the sequence (x_n) is Cauchy, since $m, n \geq N$ implies $x_m, x_n \in \text{ball}_{r_N}(x_N)$, and hence $d(x_m, x_n) < \frac{2}{N}$. Let $x = \lim x_n$. We have $x \in \overline{\text{ball}_{r_1}(x_1)} \subseteq V$, and for every n we have $x \in \overline{\text{ball}_{r_n}(x_n)} \subseteq U_n$. Thus $V \cap \bigcap U_n$ is not empty. $\qquad\square$

The reader may notice a resemblance to the proof of Theorem 1.6. Indeed, we can easily deduce the uncountability of the unit interval from the Baire category theorem. Given any sequence (t_n) in $[0,1]$, define $U_n = [0,1] \setminus \{t_n\}$; then (U_n) is a sequence of dense open sets in $[0,1]$, so its intersection must not be empty. Thus the sequence (t_n) cannot exhaust $[0,1]$.

A set is *nowhere dense* if its closure does not contain any nonempty open set. Equivalently, the complement of its closure is an open dense set. Thus, we get the following corollary to the Baire category theorem by taking complements. This is the version we will use in the sequel.

Corollary 3.34. *Let X be a complete metric space and let (C_n) be a sequence of nowhere dense subsets of X. Then $\bigcup C_n$ does not equal X.*

(In fact, the complement of $\bigcup C_n$ will be dense in X.)

Theorem 3.35. *(Approximation lemma) Let E and F be Banach spaces, let $T \in \mathcal{B}(E,F)$, and let $m > 0$ and $0 < r < 1$. Suppose that for each $y \in [F]_1$ there exists $x_0 \in E$ with $\|x_0\| \le m$ and $\|y - Tx_0\| \le r$. Then for each $y \in F$ there exists $x \in E$ with $\|x\| \le \frac{m\|y\|}{1-r}$ and $Tx = y$.*

Proof. By scaling, the hypothesis implies that for every $y \in F$ there exists $x_0 \in E$ with $\|x_0\| \le m\|y\|$ and $\|y - Tx_0\| \le r\|y\|$. Now fix $y \in F$. Without loss of generality assume $\|y\| = 1$ and find x_0 with $\|x_0\| \le m$ and $\|y - Tx_0\| \le r$. Next, applying the hypothesis to $y - Tx_0$ in place of y, find $x_1 \in E$ with $\|x_1\| \le rm$ and $\|y - T(x_0 + x_1)\| \le r^2$. Continuing in this way, we obtain a sequence (x_n) such that $\|x_n\| \le r^n m$ and $\|y - T(x_0 + \cdots + x_n)\| \le r^{n+1}$. Thus $\sum x_n$ is absolutely summable to a vector $x \in E$ of norm at most $\frac{m}{1-r}$, and $\|y - Tx\| = 0$, i.e., $y = Tx$. $\qquad\square$

In the preceding proof we used completeness of E to sum the series $\sum x_n$ in order to produce a point x that maps onto y. To see that completeness of E is necessary, just consider the inclusion map from a dense proper subspace of a Banach space into the whole space. This satisfies the hypotheses of the approximation lemma with $m = 1$ and arbitrary r, but it is not surjective.

We did not actually use completeness of F anywhere in the proof, but there is no reason to state the result in greater generality because the hypotheses can never be satisfied if E is complete but F is not. If they were then they would also be satisfied, for any strictly larger value of r, for the same map considered as a map from E into the completion of F.

But then the approximation lemma would imply that E maps onto the completion of F, which is absurd.

The Banach isomorphism theorem states that any continuous linear bijection between two Banach spaces must be an isomorphism. In other words, the inverse map is automatically continuous.

Theorem 3.36. *(Banach isomorphism theorem) Let E and F be Banach spaces and let $T \in \mathcal{B}(E, F)$ be a bijection. Then T is an isomorphism.*

Proof. Since T is surjective, $\bigcup_{n=1}^{\infty} T((E)_n) = T(E) = F$. So by Corollary 3.34, the closure of some $T((E)_n)$ must contain a nonempty open set, and by scaling this must be true of the closure of $T((E)_1)$. Let ball$_\epsilon(y)$ be an open ball contained in $\overline{T((E)_1)}$. Now any $y' \in (F)_{2\epsilon}$ is the difference of two elements of ball$_\epsilon(y)$ because we can write $y' = (y + \frac{1}{2}y') - (y - \frac{1}{2}y')$. So

$$(F)_{2\epsilon} \subseteq \text{ball}_\epsilon(y) - \text{ball}_\epsilon(y) \subseteq \overline{T((E)_2)}.$$

We conclude that $(F)_\epsilon \subseteq \overline{T((E)_1)}$. Thus the hypothesis of the approximation lemma holds with $m = \frac{1}{\epsilon}$ and arbitrary r, and it follows that $(F)_\epsilon \subseteq T((E)_1)$. That is, the image of the open unit ball of E contains a ball about the origin of F. This means that T^{-1} is bounded, so T is an isomorphism. $\square$

Our first corollary, the open mapping theorem, generalizes the Banach isomorphism theorem to surjective bounded linear maps which are not necessarily injective. Here the conclusion is not, of course, that such a map is an isomorphism, but rather that it must be *open*, meaning that $T(U)$ is open for every open $U \subseteq E$. This result can be deduced from Theorem 3.36 by factoring T through $E/\ker T$ as in Proposition 3.20 (exercise).

Theorem 3.37. *(Open mapping theorem) Let E and F be Banach spaces and let $T \in \mathcal{B}(E, F)$ be surjective. Then T is an open map.*

Now we come to the closed graph theorem. The *graph* of a map $T : E \to F$ is the set

$$\Gamma(T) = \{(x, Tx) : x \in E\} \subseteq E \oplus F.$$

We do not specify which of the norms from Definition 3.21 we are using on $E \oplus F$ because for our purposes here it doesn't matter: on a finite direct sum they are all equivalent norms.

Theorem 3.38. *(Closed graph theorem) Let E and F be Banach spaces and let $T : E \to F$ be a linear map. The following are equivalent:*

(i) T is bounded;

(ii) whenever $x_n \to 0$ in E and $Tx_n \to y$ in F we have $y = 0$;

(iii) $\Gamma(T)$ is a closed subspace of $E \oplus F$.

Proof. (i) $\Rightarrow$ (ii). If T is bounded then it is continuous (Proposition 3.12). Thus $x_n \to 0$ in E implies $Tx_n \to 0$ in F.

(ii) $\Rightarrow$ (iii). Assume (ii) and let $((x_n, Tx_n))$ be a sequence in $\Gamma(T)$ which converges to a limit point (x, y). Then $x_n - x \to 0$ in E and $T(x_n - x) \to y - Tx$ in F, so (ii) implies that $y - Tx = 0$, i.e., $Tx = y$. Thus the limit point (x, y) also belongs to $\Gamma(T)$, and we conclude that $\Gamma(T)$ is closed.

(iii) $\Rightarrow$ (i). Assume (iii). Then $\Gamma(T)$ is a closed subspace of a Banach space, so $\Gamma(T)$ is itself a Banach space. Define $P_1 : \Gamma(T) \to E$ by $P_1((x, Tx)) = x$ and $P_2 : \Gamma(T) \to F$ by $P_2((x, Tx)) = Tx$. Then P_1 is a bounded linear bijection, so Theorem 3.36 implies that P_1^{-1} is bounded, and hence that $T = P_2 \circ P_1^{-1}$ is bounded. $\square$

The Banach isomorphism theorem trivially follows from the open mapping theorem. We can also deduce it from the closed graph theorem: if $T : E \to F$ is a continuous bijection, then its graph is closed, hence the graph of T^{-1}, which is just the transpose of the graph of T, is closed, hence T^{-1} is continuous. So all three results have the same technical content, in some sense.

The most commonly used part of Theorem 3.38 is probably part (ii). A standard way to show that a map T is continuous is to check that $x_n \to x$ implies $Tx_n \to Tx$; if T is linear it is sufficient to check that $x_n \to 0$ implies $Tx_n \to 0$. The point of Theorem 3.38 (ii) is that when we verify continuity in this way we can assume that (Tx_n) converges. This means that instead of proving that (Tx_n) converges to 0 we need only show that it cannot converge to any other limit. This technique is illustrated in the proof of the next result.

Theorem 3.39. *(Principle of uniform boundedness) Let E and F be Banach spaces and let $\mathcal{S}$ be a set of bounded linear maps from E to F. Suppose that*

$$\sup_{T \in \mathcal{S}} \|Tx\| < \infty$$

for all $x \in E$. Then

$$\sup_{T \in \mathcal{S}} \|T\| < \infty.$$

Proof. We will show that $\sup \|T_n\| < \infty$ for any sequence $(T_n) \subseteq \mathcal{S}$. This is enough because if $\sup_{T \in \mathcal{S}} \|T\| = \infty$ then we could find a sequence (T_n) in $\mathcal{S}$ such that $\|T_n\| \to \infty$.

Thus, let (T_n) be a sequence in $\mathcal{S}$. Let $F_n = F$ for all $n \in \mathbf{N}$ and define $T_\infty : E \to \bigoplus^\infty F_n$ by $T_\infty x = (T_n x)$. The hypothesis on $\mathcal{S}$ tells us that for each x this sequence does belong to the l^∞ direct sum. T_∞ is clearly a linear map. To see that it is continuous, suppose $x_k \to 0$ and $T_\infty x_k \to \mathbf{y} \in \bigoplus^\infty F_n$. Then $T_\infty x_k$ must converge to $\mathbf{y}$ on each coordinate, i.e., $\lim_k T_n x_k \to y_n$ for all n. But $\lim_k T_n x_k = 0$ for all n since each T_n is bounded, so we must have $\mathbf{y} = 0$. By Theorem 3.38 (ii) we conclude that T_∞ is bounded. As $\|T_n\| \leq \|T_\infty\|$ for all n, this shows that the T_n are uniformly bounded, as desired. $\qquad \square$

There are several variations on the principle of uniform boundedness; we just mention two of them.

Corollary 3.40. *Let A be a subset of a normed vector space E. Then A is bounded iff $\sup\{|f(x)| : x \in A\} < \infty$ for all $f \in E'$.*

Proof. By Corollary 3.31, A is bounded in E if and only if $\hat{A} = \{\hat{x} : x \in A\}$ is bounded in $E'' = \mathcal{B}(E', \mathbf{F})$. By Theorem 3.39, this happens provided

$$\sup_{x \in A} |f(x)| = \sup_{x \in A} |\hat{x}(f)| < \infty$$

for all $f \in E'$. $\qquad \square$

Our last result allows us to take the pointwise limit of a sequence of bounded linear maps which is not Cauchy in norm (or *a priori* even bounded).

Theorem 3.41. *(Banach-Steinhaus theorem) Let E and F be Banach spaces and let (T_n) be a sequence in $\mathcal{B}(E, F)$. Suppose that for every $x \in E$ the sequence $(T_n x)$ converges in F. Then $\sup \|T_n\| < \infty$ and there exists $T \in \mathcal{B}(E, F)$ such that $T_n x \to Tx$ for all $x \in E$.*

Proof. Define $T : E \to F$ by setting $Tx = \lim T_n x$. Verifying that T is linear is straightforward. For each $x \in E$, the fact that the sequence $(T_n x)$ converges implies that it is bounded; so the principle of uniform boundedness yields that $M = \sup \|T_n\| < \infty$. We can finally infer that T is bounded because

$$\|Tx\| = \lim \|T_n x\| \leq M\|x\|$$

for all $x \in E$. $\qquad \square$

3.5 $C(X)$ and $C_0(X)$ spaces

In the remainder of this chapter we are going to investigate a class of Banach spaces which generalize the space c_0. (Generalizations of l^1, l^2, and l^∞ will appear in Chapter 4.) Although there are many spaces in this class, they are very well-behaved and can be understood in great detail.

Definition 3.42. For any compact Hausdorff space X, let $C(X)$ be the set of continuous functions $f : X \to \mathbf{F}$, equipped with the *sup norm*

$$\|f\|_\infty = \sup_{x \in X} |f(x)|.$$

For any locally compact Hausdorff space X, let $C_0(X)$ be the set of continuous functions $f : X \to \mathbf{F}$ with the property that for any $\epsilon > 0$ we have $|f(x)| \leq \epsilon$ outside of some compact subset $K \subseteq X$. This space is also endowed with the sup norm.

The vector space operations on $C(X)$ and $C_0(X)$ are defined pointwise, i.e., by $(af)(x) = a \cdot f(x)$ and $(f + g)(x) = f(x) + g(x)$. Of course we need to prove that af and $f + g$ are continuous whenever f and g are continuous; this can be done using the technique that we used in Theorem 2.6 to prove that sums and products of measurable functions are measurable. (See Proposition 3.45 below.)

Functions in $C_0(X)$ are said to "vanish at infinity." This is a reasonable description in cases like $C_0(\mathbf{R})$ where the requirement of being arbitrarily small off of compact sets is equivalent to saying that $\lim_{t \to \pm\infty} f(t) = 0$. It can also be interpreted literally in terms of taking the value 0 at the point ∞ in the one-point compactification of X; see Lemma 3.44 below.

As promised, c_0 is an example of a $C_0(X)$ space:

Example 3.43. The discrete topology on $\mathbf{N}$ is locally compact and Hausdorff. A subset of $\mathbf{N}$ is compact for this topology if and only if it is finite. So a function $f : \mathbf{N} \to \mathbf{F}$ vanishes at infinity if and only if $\lim_{n \to \infty} f(n) = 0$, and thus $C_0(\mathbf{N}) \cong c_0$ (Example 3.7).

Similarly, the space c defined in Exercise 3.19 is isometrically isomorphic to $C(\mathbf{N}^*)$, where $\mathbf{N}^*$ is the one-point compactification of $\mathbf{N}$ (Definition 1.49).

Every compact Hausdorff space is trivially also locally compact, and it is easy to check that in this case the definitions of $C(X)$ and $C_0(X)$ coincide. Thus $C(X)$ spaces can be considered special cases of $C_0(X)$ spaces. But there is a more fruitful observation in the opposite direction: by situating any locally compact Hausdorff space X within its one-point compactification X^* we can embed $C_0(X)$ into $C(X^*)$, and this largely reduces the theory of $C_0(X)$ spaces to the theory of $C(X)$ spaces. We describe this construction next.

Lemma 3.44. *Let X be a locally compact Hausdorff space and let X^* be its one-point compactification. Then every $f \in C_0(X)$ extends to a continuous function $\tilde{f}$ on X^* by setting $\tilde{f}(\infty) = 0$, and every function in $C(X^*)$ which takes the value 0 at ∞ restricts to a function in $C_0(X)$. The map $f \mapsto \tilde{f}$ isometrically identifies $C_0(X)$ with the set of functions in $C(X^*)$ which vanish at the point ∞.*

Proof. Let $f \in C_0(X)$. We show that the function $\tilde{f}$ defined by $\tilde{f}|_X = f$ and $\tilde{f}(\infty) = 0$ is continuous on X^*. To see this, let U be an open subset of $\mathbf{F}$; we must show that $\tilde{f}^{-1}(U)$ is open in X^*. If $0 \notin U$ then $\tilde{f}^{-1}(U) = f^{-1}(U)$ is an open subset of X, and hence an open subset of X^* (see Definition 1.49). If $0 \in U$ then there exists $\epsilon > 0$ such that $\text{ball}_\epsilon(0) \subseteq U$. By the definition of $C_0(X)$, we can find a compact subset $K \subseteq X$ such that $f(K^c) \subseteq U$. Then $\infty \in \tilde{f}^{-1}(U)$, and $X^* \setminus \tilde{f}^{-1}(U) = X \setminus f^{-1}(U)$ is a closed, and hence compact, subset of K, so again $\tilde{f}^{-1}(U)$ is open in X^*. We conclude that $\tilde{f}$ is continuous.

Now let $\tilde{f} \in C(X^*)$ be any function which satisfies $\tilde{f}(\infty) = 0$; we must show that $f = \tilde{f}|_X$ belongs to $C_0(X)$. To see this, let $\epsilon > 0$. Since $\text{ball}_\epsilon(0)$ is an open subset of $\mathbf{F}$, it follows that $\tilde{f}^{-1}(\text{ball}_\epsilon(0))$ is an open subset of X^* which contains ∞. By the definition of X^*, this implies that $K = X^* \setminus \tilde{f}^{-1}(\text{ball}_\epsilon(0))$ is a compact subset of X. Then $x \in X \setminus K$ implies $|f(x)| < \epsilon$, as desired. Finally, the map $f \mapsto \tilde{f}$ is obviously isometric since the sup norm is used in both $C_0(X)$ and $C(X^*)$. □

Proposition 3.45. *Let X be a compact Hausdorff space. Then $C(X)$ is a Banach space.*

Proof. Let $f_1, f_2 \in C(X)$; then the map $f : X \to \mathbf{F}^2$ defined by $f(x) = (f_1(x), f_2(x))$ is continuous by Corollary 1.24. Since the function $h : \mathbf{F}^2 \to \mathbf{F}$ defined by $h(s,t) = s + t$ is also continuous, it follows that the pointwise sum $f_1 + f_2 = h \circ f$ is continuous. Similarly, any scalar multiple of a continuous function is continuous because $af = k_a \circ f$ where $k_a : \mathbf{F} \to \mathbf{F}$ is defined by $k_a(t) = at$. So $C(X)$ is a vector space with operations defined pointwise.

It is straightforward to check that $\| \cdot \|_\infty$ satisfies the axioms of a norm. We verify that $C(X)$ is complete. Let (f_n) be a Cauchy sequence in $C(X)$. For each $x \in X$ we have

$$|f_m(x) - f_n(x)| \leq \sup_{y \in X} |f_m(y) - f_n(y)| = \|f_m - f_n\|_\infty,$$

so the sequence $(f_n(x))$ is Cauchy in $\mathbf{F}$. Thus we can define a function $f : X \to \mathbf{F}$ by setting $f(x) = \lim f_n(x)$.

We must show that f is continuous and that $f_n \to f$ uniformly, i.e., in sup norm. Given $\epsilon > 0$, find $N \in \mathbf{N}$ such that

$$m, n \geq N \qquad \Rightarrow \qquad \|f_m - f_n\|_\infty \leq \epsilon.$$

That is, $m, n \geq N$ implies $|f_m(x) - f_n(x)| \leq \epsilon$ for all x, and taking the limit $m \to \infty$ shows that $n \geq N$ implies $|f(x) - f_n(x)| \leq \epsilon$ for all x, i.e., $\|f - f_n\|_\infty \leq \epsilon$. Thus (f_n) converges to f uniformly.

Finally, to see that f is continuous, let $U \subseteq \mathbf{F}$ be open and let $x \in f^{-1}(U)$; we will find an open set $V \subseteq X$ such that $x \in V \subseteq f^{-1}(U)$. This is sufficient to show that the inverse image of any open set is open.

Since $f(x) \in U$, we must have $\text{ball}_\epsilon(f(x)) \subseteq U$ for some $\epsilon > 0$. Find k such that $\|f - f_k\|_\infty < \frac{\epsilon}{3}$ and let $V = f_k^{-1}(\text{ball}_{\epsilon/3}(f_k(x)))$. This set is open because f_k is continuous, and it clearly contains x. Finally, it is contained in $f^{-1}(U)$ because $f_k(y) \in \text{ball}_{\epsilon/3}(f_k(x))$ implies that

$$|f(y) - f(x)| \leq |f(y) - f_k(y)| + |f_k(y) - f_k(x)| + |f_k(x) - f(x)|$$
$$< \frac{\epsilon}{3} + \frac{\epsilon}{3} + \frac{\epsilon}{3} = \epsilon$$

and hence that $f(y) \in \text{ball}_\epsilon(f(x)) \subseteq U$, as we needed to show. $\qquad\square$

Corollary 3.46. *Let X be a locally compact Hausdorff space. Then $C_0(X)$ is a Banach space.*

Proof. $C(X^*)$ is a Banach space by Proposition 3.45. The subset $\{f \in C(X^*) : f(\infty) = 0\}$ is clearly a closed subspace, so $C_0(X)$ is a Banach space by Lemma 3.44. $\qquad\square$

The technique used in the proof of Proposition 3.45 of bounding $|f(x) - f(y)|$ by $|f(x) - g(x)| + |g(x) - g(y)| + |g(y) - f(y)|$ is called an "$\frac{\epsilon}{3}$ argument". Intuitively, if a function that is close to f takes similar values at the points x and y then f should also take similar values at x and y.

Next we look at the question of separability. In the following proof we use the fact that any continuous function f on a compact metric space is uniformly continuous, i.e., for every $\epsilon > 0$ there exists $\delta > 0$ such that

$d(x, y) < \delta$ implies $|f(x) - f(y)| < \epsilon$. To see this, suppose f is not uniformly continuous; then there exist $\epsilon > 0$ and a pair of sequences (x_n) and (y_n) such that $d(x_n, y_n) \to 0$ but $|f(x_n) - f(y_n)| \geq \epsilon$ for all n. By passing to convergent subsequences we easily derive a contradiction.

In the proof of the following theorem it will be convenient to have a special term for functions for which $d(x, y) < \delta$ implies $|f(x) - f(y)| < \epsilon$, for some particular ϵ and δ. We will call such functions ϵ-δ *continuous*.

Theorem 3.47. *Let X be a compact Hausdorff space. Then the following are equivalent:*

(i) X is second countable;
(ii) X is metrizable;
(iii) $C(X)$ is separable.

Proof. The equivalence of (i) and (ii) was proven in Corollary 1.47.

(ii) $\Rightarrow$ (iii). Suppose X is metrizable and fix a metric on X; we will construct a countable dense subset B of $C(X)$. Since X is compact it has a countable dense subset A. Observe that the collection of all functions from finite subsets of A into $\mathbf{F}_0$ is countable. Now for each finite subset $F \subseteq A$, each function $h : F \to \mathbf{F}_0$, and each rational $\epsilon > 0$ and $\delta > 0$, check whether there is a function $f \in C(X)$ which is ϵ-δ continuous and satisfies $\|f|_F - h\|_\infty < \epsilon$. If there is, choose one and include it in B. Since there are only countably many possibilities for F, h, ϵ, and δ, the set B constructed in this way will be countable.

We claim that B is dense in $C(X)$. To see this let $f \in C(X)$ and let $\epsilon > 0$; we will find $g \in B$ such that $\|f - g\|_\infty < 4\epsilon$. Without loss of generality suppose ϵ is rational. By uniform continuity there is a rational value of δ such that f is ϵ-δ continuous. Then by compactness of X we can find a finite subset F of A such that the open balls of radius δ about the points of F cover X. Next, find a function $h : F \to \mathbf{F}_0$ such that $\|f|_F - h\|_\infty < \epsilon$. By the construction of B there exists a function $g \in B$ which is ϵ-δ continuous and satisfies $\|g|_F - h\|_\infty < \epsilon$. (The fact that f satisfies these conditions ensures that some such g will have been added to B.) But then the ϵ-δ continuity of f and g together with the fact that $\|f|_F - g|_F\|_\infty < 2\epsilon$ implies that $\|f - g\|_\infty < 4\epsilon$, as claimed.

(iii) $\Rightarrow$ (i). Suppose $C(X)$ is separable and let $B \subset C(X)$ be a countable dense subset. Then let

$$\mathcal{B} = \{f^{-1}(\text{ball}_{1/2}(0)) : f \in B\}.$$

We claim that $\mathcal{B}$ is a base for the topology on X. This will show that X is second countable.

To prove the claim, let $U \subseteq X$ be any open set and let $x \in U$. We need to find $V \in \mathcal{B}$ such that $x \in V \subseteq U$. To do this, use Urysohn's lemma to find $g \in C(X)$ such that $g(x) = 0$ and $g|_{X \setminus U} = 1$, and then find $f \in \mathcal{B}$ with $\|f - g\|_\infty < \frac{1}{2}$. Let $V = f^{-1}(\text{ball}_{1/2}(0))$; this is a set in $\mathcal{B}$. We have $x \in V$ because

$$|f(x)| \leq |g(x)| + |f(x) - g(x)| < \frac{1}{2}$$

and $V \subseteq U$ because if $y \in V$ then

$$|g(y)| \leq |g(y) - f(y)| + |f(y)| < \frac{1}{2} + \frac{1}{2} = 1,$$

so that y cannot belong to $X \setminus U$. Thus V has the desired properties. $\quad\square$

In the locally compact setting metrizability is weaker than second countability (recall that any discrete space is metrizable), but second countability of X is still equivalent to separability of $C_0(X)$.

Corollary 3.48. *Let X be a locally compact Hausdorff space. Then X is second countable iff $C_0(X)$ is separable.*

Proof. X is second countable if and only if X^* is second countable by Propositions 1.51 and 1.34. X^* is second countable if and only if $C(X^*)$ is separable by Theorem 3.47. And $C(X^*)$ is separable if and only if $C_0(X)$ is separable since $C(X^*)$ is spanned by $C_0(X)$ and the function 1_{X^*}. $\quad\square$

3.6   Subalgebras

In addition to the vector space operations, $C(X)$ and $C_0(X)$ spaces also carry a pointwise product defined by $(fg)(x) = f(x)g(x)$ and, in the case of complex scalars, a pointwise complex conjugate defined by $\bar{f}(x) = \overline{f(x)}$. We can prove that continuity of f and g implies continuity of fg and $\bar{f}$ using the technique employed in the proof of Proposition 3.45 to show that af and $f + g$ are continuous. The product of two functions which vanish at infinity vanishes at infinity because the product of a function which vanishes at infinity with any bounded function will vanish at infinity.

As we noted earlier, every $C(X)$ space is also a $C_0(X)$ space, because if X is compact then the condition about vanishing at infinity becomes vacuous. So without loss of generality we will state the definitions in this section and the next only for $C_0(X)$ spaces. However, we will still treat the compact case preferentially when stating results, since most of the theorems in these sections take a nicer form in that case (and, moreover, the locally compact versions are typically proven by reducing to the compact case via Lemma 3.44).

Definition 3.49. Let X be a locally compact Hausdorff space. A *subalgebra* of $C_0(X)$ is a linear subspace $\mathcal{A}$ with the property that $f, g \in \mathcal{A}$ implies $fg \in \mathcal{A}$. It is *unital* if it contains the function 1_X which is constantly 1. It is *self-adjoint* if $f \in \mathcal{A}$ implies $\bar{f} \in \mathcal{A}$.

In this contex $\mathcal{A}$ is said to *separate points* if it is a separating family of functions, i.e., if for every distinct $x, y \in X$ there exists $f \in \mathcal{A}$ such that $f(x) \neq f(y)$.

Note that 1_X does not belong to $C_0(X)$ unless X is compact, when $C_0(X) = C(X)$. So in the noncompact case there is no possibility for a subalgebra to be unital. Also, the self-adjointness condition is vacuous when $\mathbf{F} = \mathbf{R}$ since any real-valued function f satisfies $f = \bar{f}$. In order to avoid having to constantly add some version of the phrase "which is self-adjoint in the case $\mathbf{F} = \mathbf{C}$", we will include self-adjointness as a condition regardless of whether the scalar field is $\mathbf{R}$ or $\mathbf{C}$, with the understanding that in the real case it is a vacuous condition.

Our analysis of the subalgebras of $C(X)$ is based on the Stone-Weierstrass theorem, which gives conditions under which we can conclude that a subalgebra $\mathcal{A}$ of $C(X)$ is dense. Another way to say this is that every continuous function on X is uniformly approximated by functions in $\mathcal{A}$. We prove the theorem by way of a lemma which states that the absolute value function can be uniformly approximated by polynomials. There are various ways to do this; we give an elementary argument that explicitly constructs an approximating sequence of polynomials.

Lemma 3.50. *Let $a > 0$. Then there is a sequence of polynomial functions which converges uniformly to $|t|$ on the interval $[-a, a]$.*

Proof. If $\big||t| - p(t)\big| \leq \epsilon$ on $[-1, 1]$ then $\big||\frac{t}{a}| - p\left(\frac{t}{a}\right)\big| \leq \epsilon$ on $[-a, a]$, and hence $\big||t| - ap\left(\frac{t}{a}\right)\big| \leq a\epsilon$ on $[-a, a]$. So by scaling we can assume $a = 1$.

Recursively define a sequence of polynomials p_n by setting $p_0 = 0$ and $p_{n+1}(t) = p_n(t) + \frac{1}{2}(t - p_n(t)^2)$. We will show that $p_n(t)$ increases uniformly to $\sqrt{t}$ on $[0, 1]$. This is enough because it implies that $p_n(t^2)$ increases uniformly to $|t|$ on $[-1, 1]$.

We first claim that $0 \le p_n(t) \le \sqrt{t}$ on $[0,1]$ for all n. To see this, observe that for any value of t the function $x + \frac{1}{2}(t - x^2)$ is increasing on $0 \le x \le 1$, since its derivative is positive there. So if $0 \le p_n(t) \le \sqrt{t}$ for a given value of n then

$$0 \le p_n(t) \le p_{n+1}(t) \le \sqrt{t} + \frac{1}{2}\left(t - (\sqrt{t})^2\right) = \sqrt{t}.$$

The claim follows by induction.

Now for $t \in [0,1]$ we have

$$\sqrt{t} - p_{n+1}(t) = \sqrt{t} - p_n(t) - \frac{1}{2}(t - p_n(t)^2)$$

$$= (\sqrt{t} - p_n(t))\left(1 - \frac{1}{2}(\sqrt{t} + p_n(t))\right)$$

$$\le (\sqrt{t} - p_n(t))\left(1 - \frac{\sqrt{t}}{2}\right)$$

and it inductively follows that $\sqrt{t} - p_n(t) \le (1 - \frac{\sqrt{t}}{2})^n$. So given $1 > \epsilon > 0$, if $t \le \epsilon^2$ then trivially $\sqrt{t} - p_n(t) \le \epsilon$ for any n, while $t \ge \epsilon^2$ implies $\sqrt{t} - p_n(t) \le (1 - \frac{\epsilon}{2})^n$, which is at most ϵ for sufficiently large n. We conclude that $p_n(t) \to \sqrt{t}$ uniformly on $[0,1]$, as desired. $\square$

Theorem 3.51. *(Stone-Weierstrass theorem) Let X be a compact Hausdorff space and let $\mathcal{A}$ be a closed, unital, self-adjoint subalgebra of $C(X)$ which separates points. Then $\mathcal{A} = C(X)$.*

Proof.   Consider the real case first. It follows from Lemma 3.50 that for any $f, g \in \mathcal{A}$ both

$$\max(f,g) = \frac{f + g + |f - g|}{2} \quad \text{and} \quad \min(f,g) = \frac{f + g - |f - g|}{2}$$

also belong to $\mathcal{A}$.

Next we claim that for any distinct $x, y \in X$ and any $a, b \in \mathbf{R}$ there exists $h \in \mathcal{A}$ such that $h(x) = a$ and $h(y) = b$. To see this, first find $k \in \mathcal{A}$ such that $k(x) \ne k(y)$. Then the function

$$h(z) = \frac{1}{k(y) - k(x)}\big((b - a)k(z) + ak(y) - bk(x)\big)$$

verifies the claim. (This is where we use the hypothesis that $\mathcal{A}$ is unital.)

Now let $f \in C(X)$ and $\epsilon > 0$. Fix $x \in X$ and for each $y \in X$ find $h_y \in \mathcal{A}$ such that $h_y(x) = f(x)$ and $h_y(y) = f(y)$. Then for each y we have

$$h_y(z) < f(z) + \epsilon$$

for all z in some open set V_y containing y. The open sets V_y cover X, so we can find a subcover consisting of finitely many sets $V_{y_1}, \ldots, V_{y_m}$. By the observation which began the proof we then have

$$h^x = \min(h_{y_1}, \ldots, h_{y_m}) \in \mathcal{A}.$$

This function satisfies $h^x(x) = f(x)$ and $h^x(z) < f(z) + \epsilon$ for all $z \in X$.

Do the above for each $x \in X$. Then each x has an open neighborhood U_x on which $h^x(z) > f(z) - \epsilon$, so we can find $U_{x_1}, \ldots, U_{x_n}$ which cover X and let $h = \max(h^{x_1}, \ldots, h^{x_n}) \in \mathcal{A}$. The function h satisfies

$$f(z) - \epsilon < h(z) < f(z) + \epsilon$$

for all $z \in X$, i.e., $\|f - h\|_\infty < \epsilon$, and since $\mathcal{A}$ is closed in sup norm we can now infer that $f \in \mathcal{A}$. We conclude that $\mathcal{A} = C(X)$.

In the complex case let $\mathcal{A}_{sa}$ be the set of real-valued functions in $\mathcal{A}$ and let $C(X)_{sa}$ be the set of continuous real-valued functions on X. Since $\Re f = \frac{1}{2}(f + \bar{f})$ and $\Im f = \frac{1}{2i}(f - \bar{f})$, the real and imaginary parts of any function in $\mathcal{A}$ lie in $\mathcal{A}$, and hence in $\mathcal{A}_{sa}$. It follows that $\mathcal{A}_{sa}$ is a closed unital subalgebra of $C(X)_{sa}$ that separates points, and hence $\mathcal{A}_{sa} = C(X)_{sa}$ by the real version of the theorem. Thus $\mathcal{A} = \mathcal{A}_{sa} + i \cdot \mathcal{A}_{sa} = C(X)$. $\quad\square$

Example 3.52. For any $a, b \in \mathbf{R}$, $a < b$, the polynomials are uniformly dense in $C([a, b])$.

Example 3.53. The span of the functions $\exp(2\pi i n t)$, $n \in \mathbf{Z}$ (i.e., the set of trigonometric polynomials) is uniformly dense in $C([0, 1])$.

In the following corollary we deduce a $C_0(X)$ version of the Stone-Weierstrass theorem by embedding $C_0(X)$ in $C(X^*)$. Since we can no longer ask that $\mathcal{A}$ be unital, this condition must be weakened to merely requiring that it not vanish at any point. (Incidentally, this means that the following result, when specialized to the compact case, is slightly stronger than Theorem 3.51. The condition that $\mathcal{A}$ be unital is weakened to require that it not vanish at any point.)

Corollary 3.54. *Let X be a locally compact Hausdorff space and let $\mathcal{A}$ be a closed, self-adjoint subalgebra of $C_0(X)$ which separates points. Also assume that for every $x \in X$ some $f \in \mathcal{A}$ satisfies $f(x) \neq 0$. Then $\mathcal{A} = C_0(X)$.*

Proof. As in Lemma 3.44, identify $C_0(X)$ with the functions in $C(X^*)$ which vanish at ∞. Let

$$\tilde{\mathcal{A}} = \mathcal{A} + \mathbf{F} \cdot 1_{X^*} = \{f + a \cdot 1_{X^*} : f \in \mathcal{A} \text{ and } a \in \mathbf{F}\}.$$

Then $\tilde{\mathcal{A}}$ is a self-adjoint, unital subalgebra of $C(X^*)$ which separates points. (Note that the condition that $\mathcal{A}$ not vanish at any point is needed to ensure that all the points of X are separated from ∞.) To check that $\tilde{\mathcal{A}}$ is closed, let $(f_n + a_n \cdot 1_{X^*})$ be a sequence in $\tilde{\mathcal{A}}$ which converges uniformly to $g \in C(X^*)$. Write $a = g(\infty)$ and $f = g - a \cdot 1_{X^*}$. Then

$$a_n = (f_n + a_n \cdot 1_{X^*})(\infty) \to g(\infty) = a,$$

and since $f_n + a_n \cdot 1_{X^*} \to g = f + a \cdot 1_{X^*}$ uniformly, it follows that $f_n \to f$ uniformly. Thus $f \in \mathcal{A}$, and this implies that $g \in \tilde{\mathcal{A}}$. This shows that $\tilde{\mathcal{A}}$ is closed, and Theorem 3.51 now implies that $\tilde{\mathcal{A}} = C(X)$. For any $g \in C_0(X)$, we then have $g \in \tilde{\mathcal{A}}$, so $g = f + a \cdot 1_{X^*}$ for some $f \in \mathcal{A}$ and $a \in \mathbf{F}$. But evaluating at ∞ yields $a = 0$, so $g \in \mathcal{A}$. We conclude that $\mathcal{A} = C_0(X)$. $\square$

We now proceed to characterize the closed, unital, self-adjoint subalgebras of $C(X)$. Recall that an equivalence relation $\sim$ on a topological space X is *closed* if the set $\{(x, y) : x \sim y\}$ is a closed subset of $X \times X$, and that a function $f \in C(X)$ *respects* $\sim$ if $x \sim y$ implies $f(x) = f(y)$.

If $\sim$ is any equivalence relation on a compact Hausdorff space X, it is routine to check that the set $\mathcal{A}_\sim$ of all continuous scalar-valued functions on X which respect $\sim$ is a closed, unital, self-adjoint subalgebra of $C(X)$. Conversely, if $\mathcal{A}$ is any family of functions in $C(X)$, the relation $\sim_\mathcal{A}$ defined by setting $x \sim_\mathcal{A} y$ if $f(x) = f(y)$ for all $f \in \mathcal{A}$ is a closed equivalence relation (exercise).

Theorem 3.55. *Let X be a compact Hausdorff space. Then any closed, unital, self-adjoint subalgebra $\mathcal{A}$ of $C(X)$ is isometrically isomorphic to $C(X/\sim_\mathcal{A})$. The map taking $\sim$ to $\mathcal{A}_\sim$ describes a bijection between the closed equivalence relations on X and the closed, unital, self-adjoint subalgebras of $C(X)$, and its inverse is the map taking $\mathcal{A}$ to $\sim_\mathcal{A}$.*

Proof. Let $\mathcal{A}$ be a closed, unital, self-adjoint subalgebra of $C(X)$. By Theorem 1.45 the quotient space $X/\sim_\mathcal{A}$ is compact Hausdorff, so for $f \in \mathcal{A}$ define $\tilde{f} \in C(X/\sim_\mathcal{A})$ as in Proposition 1.25 and let $\tilde{\mathcal{A}}$ be the set of all such functions $\tilde{f}$ for $f \in \mathcal{A}$. It is straightforward to check that $\tilde{\mathcal{A}}$ is a self-adjoint unital subalgebra of $C(X/\sim_\mathcal{A})$. We clearly have $\|\tilde{f}\|_\infty = \|f\|_\infty$ for all f, so that $\tilde{\mathcal{A}}$ is isometric to $\mathcal{A}$, and since $\mathcal{A}$ is complete this shows that $\tilde{\mathcal{A}}$ is closed. Moreover, $\tilde{\mathcal{A}}$ separates points by the definition of $\sim_\mathcal{A}$. So the Stone-Weierstrass theorem yields that $\tilde{\mathcal{A}} = C(X/\sim_\mathcal{A})$. We conclude that $\mathcal{A}$ is isometrically isomorphic to $C(X/\sim_\mathcal{A})$.

We show that the maps taking $\sim$ to $\mathcal{A}_\sim$ and taking $\mathcal{A}$ to $\sim_\mathcal{A}$ are each other's inverses. First, let $\sim$ be a closed equivalence relation and let $\mathcal{A}_\sim$ be the corresponding subalgebra of $C(X)$. We want to show that the equivalence relation derived from $\mathcal{A}_\sim$ equals $\sim$. It is clear that $x \sim y$ implies that $f(x) = f(y)$ for all $f \in \mathcal{A}_\sim$. Conversely, if $x \not\sim y$ then by Urysohn's lemma we can find a function $g \in C(X/\sim)$ such that $g([x]) \neq g([y])$. Then $g \circ \pi$, where $\pi : X \to X/\sim$ is the quotient map, is a continuous function on X which respects $\sim$ and separates x and y. Thus there is a function in $\mathcal{A}_\sim$ which takes different values on x and y, which is what we needed.

Next, let $\mathcal{A}$ be a closed, unital, self-adjoint subalgebra of $C(X)$ and let $\sim_\mathcal{A}$ be the corresponding equivalence relation on X. We want to show that the subalgebra derived from $\sim_\mathcal{A}$ equals $\mathcal{A}$. It is clear that every function in $\mathcal{A}$ respects $\sim_\mathcal{A}$. To prove the converse, observe that if f is any continuous function on X that respects $\sim$ then, in the notation of the first part of the proof, $\tilde{f}$ belongs to $\tilde{\mathcal{A}}$ (Proposition 1.25), and hence f belongs to $\mathcal{A}$. $\quad\square$

We emphasize the remarkable fact we have just proven: if X is a compact Hausdorff space then any closed, unital, self-adjoint subalgebra of $C(X)$ is isometrically isomorphic to some $C(Y)$.

The closed, self-adjoint subalgebras of $C_0(X)$ for locally compact X can be classified in a similar way. This can be done by passing to $C(X^*)$ and adding the constant functions to the subalgebra, and then invoking Theorem 3.55. We leave the details to the exercises.

We can use Theorem 3.55 to give a quick proof that every quotient of a compact metrizable space by a closed equivalence relation is metrizable. This is not so easy to show directly, because in general there is no "natural" metric on the quotient space (cf. Exercise 3.33).

Corollary 3.56. *Let $\sim$ be a closed equivalence relation on a compact metrizable space X. Then the quotient space $X/\sim$ is metrizable.*

Proof. We know that $C(X/\sim)$ is isometric to the set of functions in $C(X)$ which respect $\sim$. Since X is metrizable, $C(X)$ is separable by Theorem 3.47. By Corollary 1.35 (b), this implies that $C(X/\sim)$ is also separable, so by Theorem 3.47 again we conclude that $X/\sim$ is metrizable. $\quad\square$

3.7 Ideals and homomorphisms

We saw in the last section that subalgebras of $C(X)$ correspond to quotients of X. We will now show that quotients of $C(X)$ correspond to subsets of X.

Definition 3.57. Let X be a locally compact Hausdorff space. An *ideal* of $C_0(X)$ is a subspace $\mathcal{I}$ with the property that $f \in \mathcal{I}$ and $g \in C_0(X)$ implies $fg \in \mathcal{I}$.

The point of this definition is that if $\mathcal{I}$ is an ideal then the product operation on $C_0(X)$ descends to the quotient space $C_0(X)/\mathcal{I}$. We need the product not to depend on the choice of representatives, i.e., $f - f' \in \mathcal{I}$ and $g - g' \in \mathcal{I}$ must imply $fg - f'g' \in \mathcal{I}$. This works because

$$fg - f'g' = f(g - g') + (f - f')g' \in \mathcal{I}.$$

To see that the defining condition for ideals is necessary, take $f \in \mathcal{I}$, $f' = 0$, $g \in C_0(X)$, and $g' = g$.

Whereas the Stone-Weierstrass theorem was the key result in our analysis of the subalgebras of $C(X)$, our treatment of quotients hinges on the Tietze extension theorem. We will prove this theorem using the approximation lemma.

The Tietze extension theorem states that any continuous function defined on a closed subset of a compact Hausdorff space can be extended to a continuous function on the whole space. The relevance of this result to Banach spaces may not be immediately apparent. But given a closed subset K of a compact Hausdorff space X, consider the restriction map from $C(X)$ to $C(K)$ which takes a function on X to its restriction to K. This is a bounded linear map, and the main content of the Tietze extension theorem is just the fact that this map is surjective.

Theorem 3.58. *(Tietze extension theorem) Let X be a compact Hausdorff space and let K be a closed subset of X. Then for any $f \in C(K)$ there exists $\tilde{f} \in C(X)$ such that $\tilde{f}|_K = f$ and $\|\tilde{f}\|_\infty = \|f\|_\infty$.*

Proof. First consider the real case. Define $T : C(X) \to C(K)$ by $Tf = f|_K$. This is a bounded linear map. We show that T is surjective by verifying the hypothesis of the approximation lemma (Theorem 3.35) with $m = \frac{1}{3}$ and $r = \frac{2}{3}$. For any $g \in [C(K)]_1$, let $C = g^{-1}\left(\left[-1, -\frac{1}{3}\right]\right)$ and $D = g^{-1}\left(\left[\frac{1}{3}, 1\right]\right)$. By Urysohn's lemma there exists a continuous function $f_0 : X \to \left[-\frac{1}{3}, \frac{1}{3}\right]$ such that $f_0|_C = -\frac{1}{3}$ and $f_0|_D = \frac{1}{3}$. Then $\|f_0\|_\infty = \frac{1}{3}$ and $\|g - Tf_0\|_\infty \leq \frac{2}{3}$. The approximation lemma now implies that for any $f \in C(K)$ there exists $\tilde{f} \in C(X)$ such that $\|\tilde{f}\|_\infty \leq \|f\|$ and $T\tilde{f} = f$, i.e., $\tilde{f}|_K = f$. This completes the proof in the real case.

In the case of complex scalars, by separating into real and complex parts and extending each part separately, for any $f \in C(K)$ we can find

$f' \in C(X)$ such that $f'|_K = f$. However, $\|f'\|_\infty$ could be as large as $\sqrt{2} \cdot \|f\|_\infty$. To remedy this defect let $a = \|f\|_\infty$ and consider the map $h : \mathbf{C} \to \mathbf{C}$ defined by

$$h(z) = \begin{cases} z & \text{if } |z| \le a \\ \frac{a}{|z|}z & \text{if } |z| > a. \end{cases}$$

This map is continuous, its range is the closed disk $\{z : |z| \le a\}$, and it fixes every element of this disk. Thus $\tilde{f} = h \circ f'$ belongs to $C(X)$ and satisfies $\tilde{f}|_K = f$ and $\|\tilde{f}\|_\infty = \|f\|_\infty$. $\qquad\square$

We proceed to characterize ideals and quotients of $C(X)$ spaces. If K is any subset of a compact Hausdorff space X, it is routine to check that the set $\mathcal{I}_K$ of all continuous scalar-valued functions on X which vanish on K is a closed, self-adjoint ideal of $C(X)$. Conversely, if $\mathcal{I}$ is any family of functions on $C(X)$, the set $K_{\mathcal{I}}$ of $x \in X$ on which every function in $\mathcal{I}$ vanishes is a closed subset (exercise).

Lemma 3.59. *Let X be a compact Hausdorff space and let $\mathcal{I}$ be a closed, self-adjoint ideal of $C(X)$. Then every continuous function on X that vanishes on $K_{\mathcal{I}}$ belongs to $\mathcal{I}$.*

Proof. To avoid triviality we can assume that $\mathcal{I} \ne C(X)$. We start by showing that $K_{\mathcal{I}}$ must be nonempty. Otherwise, for every $x \in X$ there exists $f_x \in \mathcal{I}$ such that $f_x(x) \ne 0$. Define $U_x = f_x^{-1}(\mathbf{F} \setminus \{0\})$; then the family $\{U_x : x \in X\}$ is an open cover of X, so we can find a finite subcover $\{U_{x_1}, \ldots, U_{x_n}\}$. Then the function

$$f = |f_{x_1}|^2 + \cdots + |f_{x_n}|^2$$

is strictly positive on X, and it belongs $\mathcal{I}$ since $|f_{x_i}|^2 = f_{x_i}\bar{f}_{x_i}$ for all i. It follows that $\frac{1}{f}$ belongs to $C(X)$, and hence that $g = f \cdot \frac{g}{f} \in \mathcal{I}$ for every $g \in C(X)$, which shows that $\mathcal{I} = C(X)$. We conclude that if $\mathcal{I} \ne C(X)$ then $K_{\mathcal{I}}$ must be nonempty.

Now define

$$\tilde{\mathcal{I}} = \mathcal{I} + \mathbf{F} \cdot 1_X = \{f + a \cdot 1_X : f \in \mathcal{I} \text{ and } a \in \mathbf{F}\}.$$

It is clear that $\tilde{\mathcal{I}}$ is a unital, self-adjoint subalgebra of $C(X)$, and the argument used in the proof of Corollary 3.54 to prove that $\tilde{\mathcal{A}}$ is closed shows that $\tilde{\mathcal{I}}$ is closed. (The role of ∞ can now be played by any $x \in K_{\mathcal{I}}$.) So according to Theorem 3.55,

$$\tilde{\mathcal{I}} = \{f \in C(X) : x \sim_{\tilde{\mathcal{I}}} y \text{ implies } f(x) = f(y)\}.$$

Clearly, if $x, y \in K_{\mathcal{I}}$ then $x \sim_{\tilde{\mathcal{I}}} y$. But if $x \notin K_{\mathcal{I}}$ then x is equivalent only to itself. This is because there exists $f \in \mathcal{I}$ such that $f(x) \neq 0$, and for any $y \neq x$ Urysohn's lemma provides a function $g \in C(X)$ such that $g(x) = 1$ and $g(y) = 0$; then $fg(x) \neq 0 = fg(y)$, and $fg \in \mathcal{I} \subseteq \tilde{\mathcal{I}}$, so x is not equivalent to y. It follows that any continuous function that vanishes on $K_{\mathcal{I}}$ respects $\sim_{\tilde{\mathcal{I}}}$, and must therefore lie in $\tilde{\mathcal{I}}$, and hence (since it is zero on $K_{\mathcal{I}}$) in $\mathcal{I}$. This proves the lemma. $\square$

Theorem 3.60. *Let X be a compact Hausdorff space. Then the quotient of $C(X)$ by any closed, self-adjoint ideal $\mathcal{I}$ is isometrically isomorphic to $C(K_{\mathcal{I}})$. The map taking K to $\mathcal{I}_K$ describes a bijection between the closed subsets of X and the closed, self-adjoint ideals of $C(X)$, and its inverse is the map taking $\mathcal{I}$ to $K_{\mathcal{I}}$.*

Proof. Let $\mathcal{I}$ be a closed, self-adjoint ideal of $C(X)$ and define $T : C(X) \to C(K_{\mathcal{I}})$ by $Tf = f|_{K_{\mathcal{I}}}$. This map is surjective by the Tietze extension theorem, and its kernel is $\mathcal{I}$ by the lemma. Thus T induces a vector space isomorphism $\tilde{T} : C(X)/\mathcal{I} \to C(K_{\mathcal{I}})$. It is clear that $\|T\| \leq 1$, so by Proposition 3.20 $\|\tilde{T}\| \leq 1$ too. To verify that $\tilde{T}$ is an isometry, we must show that $\|f + \mathcal{I}\| \leq \|f|_{K_{\mathcal{I}}}\|_{\infty}$ for all $f \in C(X)$. But given $f \in C(X)$, by the Tietze extension theorem we can find $\tilde{f} \in C(X)$ such that $\tilde{f}|_{K_{\mathcal{I}}} = f|_{K_{\mathcal{I}}}$ and $\|\tilde{f}\|_{\infty} = \|f|_{K_{\mathcal{I}}}\|_{\infty}$. Then $\tilde{f} \in f + \mathcal{I}$ because $f - \tilde{f}$ vanishes on $K_{\mathcal{I}}$, so that

$$\|f + \mathcal{I}\| \leq \|\tilde{f}\|_{\infty} = \|f|_{K_{\mathcal{I}}}\|_{\infty},$$

as desired. This completes the proof of the first statement.

We show that the maps taking K to $\mathcal{I}_K$ and taking $\mathcal{I}$ to $K_{\mathcal{I}}$ are each other's inverses. First, let K be a closed subset of X and let $\mathcal{I}_K$ be the corresponding ideal of $C(X)$. We want to show that the closed set derived from $\mathcal{I}_K$ equals K. It is clear that $x \in K$ implies that $f(x) = 0$ for all $f \in \mathcal{I}_K$. Conversely, we need to show that if $x \notin K$ then there is some function in $\mathcal{I}_K$ which does not vanish on x. But since any continuous function that vanishes on K belongs to $\mathcal{I}_K$, this follows from Urysohn's lemma.

Next, let $\mathcal{I}$ be a closed, self-adjoint ideal of $C(X)$ and let $K_{\mathcal{I}}$ be the corresponding subset of X. We want to show that the ideal derived from $K_{\mathcal{I}}$ equals $\mathcal{I}$. But it is clear that every function in $\mathcal{I}$ vanishes on $K_{\mathcal{I}}$, and the converse statement was proven in the lemma. $\square$

This is the second remarkable fact about the algebraic structure of $C(X)$ spaces. If X is a compact Hausdorff space then the quotient of $C(X)$ by any closed, self-adjoint ideal is isometrically isomorphic to some $C(Y)$.

Now that we have characterized closed, self-adjoint ideals, it is easy to analyze bounded linear maps that respect algebraic structure. The relevant definition is the following.

Definition 3.61. Let X and Y be locally compact Hausdorff spaces. A *homomorphism* from $C_0(X)$ to $C_0(Y)$ is a linear map $T : C_0(X) \to C_0(Y)$ which satisfies the law $T(fg) = T(f)T(g)$ for all $f, g \in C_0(X)$. If it also satisfies $T(\bar{f}) = \overline{T(f)}$ for all $f \in C_0(X)$ then it is *self-adjoint*. It is *unital* if $T(1_X) = 1_Y$.

As usual, the condition that $T(1_X) = 1_Y$ is only relevant in the compact case. But we can always ensure this condition by passing to X^* and Y^*; if $T : C_0(X) \to C_0(Y)$ is a (bounded, self-adjoint) homomorphism then its extension $\tilde{T} : C(X^*) \to C(Y^*)$ defined by setting $\tilde{T}(1_{X^*}) = 1_{Y^*}$ is also a (bounded, self-adjoint) homomorphism.

We start with homomorphisms into $\mathbf{F}$. For any point x in a compact Hausdorff space X, define the *point evaluation* $\hat{x} : C(X) \to \mathbf{F}$ by $\hat{x}(f) = f(x)$. It is routine to verify that $\hat{x}$ is a bounded, unital, self-adjoint homomorphism (exercise). Let $\hat{X} = \{\hat{x} : x \in X\}$ and equip it with the weak topology determined by the family of functions $\hat{f} : \hat{x} \to f(x)$ for $f \in C(X)$.

Corollary 3.62. *Let X be a compact Hausdorff space. Then every bounded, unital, self-adjoint homomorphism from $C(X)$ to $\mathbf{F}$ equals $\hat{x}$ for some $x \in X$. The map taking x to $\hat{x}$ is a homeomorphism between X and $\hat{X}$.*

Proof. Let $T : C(X) \to \mathbf{F}$ be a bounded, unital, self-adjoint homomorphism. Then $\mathcal{I} = \ker(T)$ is a closed, self-adjoint ideal of $C(X)$, and by Theorem 3.60 we have $\mathcal{I} = \mathcal{I}_K$ and $C(X)/\mathcal{I} \cong C(K)$ for some closed subset K of X. But T is surjective (since it is unital), so we also have $C(X)/\mathcal{I} \cong \mathbf{F}$. Thus $C(K) \cong \mathbf{F}$, and this shows that K must contain exactly one point, say $K = \{x\}$. For any $f \in C(X)$ we then have $f - a \cdot 1_X \in \mathcal{I}_K = \ker(T)$ where $a = f(x)$, and hence

$$T(f) = T(a \cdot 1_X) = a = f(x) = \hat{x}(f).$$

This shows that $T = \hat{x}$.

For the second statement, let $\phi : X \to \hat{X}$ be the map defined by $\phi(x) = \hat{x}$. This map is injective because if $x \neq y$ then there is a function $f \in C(X)$ which separates x and y, which implies that $\hat{x}(f) = f(x) \neq f(y) = \hat{y}(f)$. Next, since the topology on $\hat{X}$ is the weak topology determined by the family of functions $\hat{f}$ for $f \in C(X)$, to prove that ϕ is continuous it will suffice to show that $\hat{f} \circ \phi : X \to \mathbf{F}$ is continuous for all $f \in C(X)$. But

$$(\hat{f} \circ \phi)(x) = \hat{f}(\phi(x)) = \hat{f}(\hat{x}) = f(x),$$

so $\hat{f} \circ \phi = f$. Thus $\hat{f} \circ \phi$ is continuous, and we conclude that ϕ is continuous. To see that $\hat{X}$ is Hausdorff, let $\hat{x}, \hat{y} \in \hat{X}$ be distinct. Then we can find $f \in C(X)$ such that $\hat{f}(\hat{x}) = f(x) = 0$ and $\hat{f}(\hat{y}) = f(y) = 1$, and $\hat{f}^{-1}(\text{ball}_{1/2}(0))$ and $\hat{f}^{-1}(\text{ball}_{1/2}(1))$ will be disjoint open subsets of $\hat{X}$ which respectively contain $\hat{x}$ and $\hat{y}$. So ϕ is a homeomorphism by Corollary 1.41. $\qquad\square$

This result shows that X can be "recovered" from $C(X)$ as, up to homeomorphism, the set of bounded, unital, self-adjoint homomorphisms from $C(X)$ into $\mathbf{F}$, equipped with the weak topology determined by evaluating on elements of $C(X)$.

In the locally compact case, we cannot require homomorphisms to be unital, so we instead define $\hat{X}$ to be the set of nonzero bounded, self-adjoint homomorphisms from $C_0(X)$ into $\mathbf{F}$. We still get that $\hat{X}$ is homeomorphic to X (exercise). Including the zero homomorphism in $\hat{X}$ corresponds to forming the one-point compactification of X.

The characterization of homomorphisms between arbitrary $C(X)$ and $C(Y)$ follows fairly easily.

Corollary 3.63. *Let X and Y be compact Hausdorff spaces. If $\phi : Y \to X$ is continuous then the map $f \mapsto f \circ \phi$ is a bounded, unital, self-adjoint homomorphism from $C(X)$ to $C(Y)$. Every bounded, unital, self-adjoint homomorphism from $C(X)$ to $C(Y)$ has this form.*

Proof. Verification of the first statement is routine. For the second, let $T : C(X) \to C(Y)$ be a bounded, unital, self-adjoint homomorphism. Then composition with T takes bounded, unital, self-adjoint homomorphisms from $C(Y)$ into $\mathbf{F}$ to bounded, unital, self-adjoint homomorphisms from $C(X)$ into $\mathbf{F}$. In other words, it defines a map $\hat{\phi}$ from $\hat{Y}$ to $\hat{X}$.

We check that $\hat{\phi}$ is continuous. Since the topology on $\hat{X}$ is the weak topology determined by the functions $\hat{f}$ for $f \in C(X)$, it will suffice to show that $\hat{f} \circ \hat{\phi} : \hat{Y} \to \mathbf{F}$ is continuous for every $f \in C(X)$. But

$$(\hat{f} \circ \hat{\phi})(\hat{y}) = \hat{f}(\hat{\phi}(\hat{y})) = (\hat{\phi}(\hat{y}))(f) = \hat{y}(Tf) = (Tf)(y);$$

since $Tf \in C(Y)$ and the map $y \mapsto \hat{y}$ is a homeomorphism between Y and $\hat{Y}$, this shows that $\hat{f} \circ \hat{\phi}$ is indeed continuous.

We now define a map $\phi : Y \to X$ by setting $\phi(y) = x$ when $\hat{\phi}(\hat{y}) = \hat{x}$. Since $\hat{\phi}$ is continuous, so is ϕ. For any $f \in C(X)$ and $y \in Y$ we have

$$(f \circ \phi)(y) = f(\phi(y)) = \hat{\phi}(\hat{y})(f) = \hat{y}(Tf) = Tf(y);$$

this shows that $f \circ \phi = Tf$. So T does have the desired form. □

3.8 Exercises

3.1. Let E be a normed vector space. Show that the map $(a, x) \mapsto ax$ is continuous from $\mathbf{F} \times E$ to E and the map $(x, y) \mapsto x + y$ is continuous from $E \times E$ to E, using the product topology in both cases.

3.2. Prove that $l^1(\mathbf{N})$ and $l^1(\mathbf{Z})$ are isometrically isomorphic. What about $l^1(\mathbf{N}^2)$?

3.3. Show that $l^1 \subset l^2 \subset l^\infty$ and that both containments are proper.

3.4. Carry out the proof that l^2 is complete.

3.5. Why can't we apply Lemma 3.11 to the sequence $(\mathbf{e}^n)$ (introduced just before Theorem 3.25) to show that l^∞ is separable?

3.6. Prove that every separable metric space embeds isometrically in l^∞.

3.7. Is the restriction of the l^2 norm to $l^1 \subset l^2$ equivalent to the l^1 norm? Same question for l_n^2 and l_n^1.

3.8. Verify that both formulas for $\|T\|$ in Definition 3.14 produce the same result.

3.9. Prove Proposition 3.15.

3.10. Let E_0 be a subspace of a normed vector space E. Prove that its closure $\overline{E}_0$ is a closed subspace of E.

3.11. Prove that any two norms on $\mathbf{F}^n$ are equivalent. Prove that any finite-dimensional subspace of a normed vector space is automatically closed.

3.12. Let E be an infinite-dimensional normed vector space. Use Proposition 3.19 and Exercise 3.11 to construct a sequence (x_n) in E such that $\|x_n\| = 1$ for all n and $\|x_m - x_n\| \geq \frac{1}{2}$ for all $m \neq n$. Deduce that the norm topology on E is not locally compact.

3.13. Let E be a normed vector space and E_0 a closed subspace. Is the norm topology on E/E_0 coming from the quotient norm the same as the quotient topology coming from the norm topology on E?

3.14. Let E be a normed vector space and E_0 a closed subspace. Prove that the image of $(E)_1$ under the projection map $\pi : E \to E/E_0$ is $(E/E_0)_1$.

3.15. Prove that the map $\tilde{T}$ in Proposition 3.20 is an isometry if and only if $\|y\| = \inf\{\|x\| : Tx = y\}$ for all $y \in T(E)$.

3.16. Prove that $\| \cdot \|_2$ is a norm on the l^2 direct sum of any countable family of normed vector spaces.

3.17. Let f be a linear functional on a normed vector space E. Prove that f is bounded if and only if $\ker f$ is closed. (Use Proposition 3.19 and the fact that all norms on $\mathbf{F}$ are equivalent.)

3.18. Prove Theorem 3.25 (ii).

3.19. Let c be the subspace of l^∞ consisting of all convergent sequences of scalars. Prove that $c' \cong l^1$.

3.20. Let E be a normed vector space, let E_0 be a closed subspace, and let $x \in E \setminus E_0$. Prove that $\text{span}(E_0 \cup \{x\})$ is closed. (Apply Corollary 3.29 to a Cauchy sequence in this span.)

3.21. Prove that every separable Banach space is isometrically isomorphic to a closed subspace of l^∞. (Use the Hahn-Banach theorem.)

3.22. Let E be a separable Banach space. Prove that there is a bounded linear map from l^1 onto E. Prove that E is isometrically isomorphic to a quotient of l^1. (Use Exercise 3.15.)

3.23. Let E be a separable normed vector space, E_0 a subspace of E, and $f : E_0 \to l^\infty$ a bounded linear map. Prove that f extends to a bounded linear map $\tilde{f} : E \to l^\infty$ such that $\|\tilde{f}\| = \|f\|$.

3.24. Prove the open mapping theorem (Theorem 3.37). (Use the Banach isomorphism theorem and Exercise 3.14.)

3.25. Let $\mathbf{a} \in l^1$. Prove that not every element of l^1 is of the form $(a_n b_n)$ for some $\mathbf{b} \in l^\infty$. (Consider the map from l^∞ to l^1 taking $\mathbf{b}$ to $(a_n b_n)$.)

3.26. Let $(\mathbf{a}^n)$ be a sequence in l^2 and suppose that for every $\mathbf{b} \in l^2$ the scalar sequence $(\langle \mathbf{a}^n, \mathbf{b} \rangle)$ belongs to l^2. Prove that the map taking $\mathbf{b}$ to this sequence is a bounded linear map from l^2 to itself.

3.27. Let E and F be Banach spaces and let $\mathcal{S}$ be a set of bounded linear maps from E to F. Suppose that $\sup_{T \in \mathcal{S}} |f(Tx)| < \infty$ for all $x \in E$ and all $f \in F'$. Prove that $\sup_{T \in \mathcal{S}} \|T\| < \infty$.

3.28. Prove that the closure of a subalgebra of any $C(X)$ is still a subalgebra.

3.29. Let K be the Cantor set. Prove that the continuous functions from K into $\mathbf{F}$ whose range is finite are dense in $C(K)$.

3.30. Let X and Y be compact Hausdorff spaces. For each $f \in C(X)$ and $g \in C(Y)$ let $f \otimes g \in C(X \times Y)$ be the function $(f \otimes g)(x,y) = f(x)g(y)$, and let

$$C(X) \otimes C(Y) = \overline{\text{span}}\{f \otimes g : f \in C(X) \text{ and } g \in C(Y)\}$$

(taking the closure in sup norm). Show that $C(X) \otimes C(Y) = C(X \times Y)$.

3.31. Prove that if $\mathcal{A}$ is a family of functions in any $C(X)$, the relation $\sim_{\mathcal{A}}$ defined by setting $x \sim_{\mathcal{A}} y$ if $f(x) = f(y)$ for all $f \in \mathcal{A}$ is a closed equivalence relation. (Consider the functions $f(x) - f(y)$ on $X \times X$.)

3.32. Let X be a locally compact Hausdorff space and let $\mathcal{A}$ be a closed, self-adjoint subalgebra of $C_0(X)$. Prove that there is a closed equivalence relation $\sim$ on X^* such that $\mathcal{A}$ consists of precisely those functions in $C_0(X) \subset C(X^*)$ which respect $\sim$. Is every closed equivalence relation on X the restriction of a closed equivalence relation on X^*?

3.33. Given an equivalence relation $\sim$ on a metric space X, define

$$\tilde{d}([x], [y]) = \inf\{d(x_1, y_1) + \cdots + d(x_n, y_n)\},$$

taking the infimum over all finite sequences satisfying $x_1 \sim x$, $y_n \sim y$, and $y_i \sim x_{i+1}$ for $1 \leq i < n$. Prove that this defines a pseudometric on the quotient set $X/\sim$. Find a compact metric space and a closed equivalence relation for which this pseudometric is not a metric.

3.34. Prove or disprove: if X is locally compact Hausdorff and K is a closed subset of X, any function in $C_0(K)$ can be extended to one in $C_0(X)$.

3.35. Prove that if $\mathcal{I}$ is a family of functions in any $C(X)$, the set $K_{\mathcal{I}}$ of points in X on which every function in $\mathcal{I}$ vanishes is a closed subset.

3.36. For any point x in a locally compact Hausdorff space X, verify that $\hat{x} : f \mapsto f(x)$ is a bounded, self-adjoint homomorphism from $C_0(X)$ into $\mathbf{F}$.

3.37. Let X be a locally compact Hausdorff space. Classify the closed, self-adjoint ideals of $C_0(X)$.

3.38. Let X be a locally compact Hausdorff space. Prove that every nonzero bounded self-adjoint homomorphism from $C_0(X)$ into $\mathbf{F}$ equals $\hat{x}$, defined as in Exercise 3.36, for some $x \in X$. Prove that the set of all such homomorphisms is homeomorphic to X when given the weak topology determined by evaluating at elements of $C_0(X)$.

3.39. Let X be a compact Hausdorff space. Prove that every closed ideal $\mathcal{I}$ of $C(X)$ is self-adjoint. (Given $f \in \mathcal{I}$, prove that the functions $\frac{n|f|^2}{1+n|f|^2}\bar{f}$ converge uniformly to $\bar{f}$.)

3.40. Let X and Y be compact Hausdorff spaces and let $T : C(X) \to C(Y)$ be a unital, self-adjoint homomorphism. Prove that T is bounded and $\|T\| = 1$. (If f is real-valued and $a > \|f\|_\infty$ or $a < -\|f\|_\infty$, then $f - a \cdot 1_X$ is invertible in $C(X)$. If f is complex-valued, then $\|f\|_\infty = \| |f|^2 \|_\infty^{1/2}$.)

Chapter 4

Dual Banach Spaces

4.1 Weak* topologies

The dual of a separable Banach space may be separable (like $l^1 \cong (c_0)'$) or nonseparable (like $l^\infty \cong (l^1)'$). In the latter case it becomes somewhat hard to handle as a Banach space. However, it carries another topology, the weak* topology, which is quite manageable, at least on bounded subsets.

Recall from Definition 3.30 that if E is a normed vector space and $x \in E$ then the evaluation functional $\hat{x} : E' \to \mathbf{F}$ is defined by $\hat{x}(f) = f(x)$.

Definition 4.1. Let E be a normed vector space. The *weak* topology* on its dual space E' is the weak topology determined by the family of maps $\hat{x} : E' \to \mathbf{F}$ for $x \in E$.

This was the topology we used on $\hat{X} \subset C(X)'$ in Section 3.7. In terms of sequences, we have $f_n \to f$ weak* in E' if and only if $f_n(x) \to f(x)$ for all $x \in E$. Every weak* neighborhood of 0 contains a set of the form $\{f : |f(x_i)| < \epsilon \text{ for } 1 \leq i \leq n\}$ for some $\epsilon > 0$ and $x_1, \ldots, x_n \in E$. Thus the translations of these sets constitute a basis for the weak* topology.

Observe that if $\|f - f_n\| \to 0$ then $f_n(x) \to f(x)$ for any x. This shows that every $\hat{x}$ is continuous for the norm topology. Therefore the weak* topology is always weaker than the norm topology. (Every weak* open/closed set is also norm open/closed.)

If $E' \cong F$ then we call E a *predual* of F. Not "the" predual, because if E_0 is a dense subspace of E then $E_0' \cong E'$ by Proposition 3.16, so there could be many normed vector spaces with the same dual space up to isometric isomorphism. Even if we restrict attention to Banach spaces, Theorem 3.25 (i) and Exercise 3.19 give examples of nonisometric spaces whose duals are isometrically isomorphic. (We will see in Section 4.4 that c and c_0 cannot

119

be isometrically isomorphic because the closed unit ball of the former has extreme points, while the closed unit ball of the latter does not.) The point is that the weak* topology on a dual Banach space is not well-defined until we specify which predual we have in mind. Usually there is a canonical choice, though.

Weak* topologies bear some resemblance to product topologies. In the following example that resemblance is actually an identity.

Example 4.2. Let $l_0^1 \subset l^1$ be the dense subspace consisting of all scalar sequences which are eventually zero. Equivalently, it is the linear span of the sequences $\mathbf{e}^n$ introduced just before Theorem 3.25. By Proposition 3.16 and Theorem 3.25 (b) we have $(l_0^1)' \cong l^\infty$.

Since sums and scalar multiples of continuous functions are automatically continuous, the weak topology on l^∞ determined by evaluation on the elements $\mathbf{e}^n$ equals the weak topology determined by evaluation on everything in their span, l_0^1. But if we regard l^∞ as a subset of $\mathbf{F}^{\mathbf{N}}$, the product of a sequence of copies of $\mathbf{F}$, then evaluation on $\mathbf{e}^n$ is just the restriction of the nth coordinate projection. So the weak* topology on $l^\infty \cong (l_0^1)'$ is the restriction to l^∞ of the product topology on $\mathbf{F}^{\mathbf{N}}$.

In general, if $A \subseteq E$ is any subset which spans E then the weak* topology on E' is also determined by the maps $\hat{x}$ with x restricted to range over A. However, "span" cannot be replaced by "closed span" in this comment. If E_0 is a proper dense subspace of E then $E_0' \cong E'$, but the weak* topologies corresponding to the two preduals are not the same. This will follow from Corollary 4.8 below.

Nonetheless, if E_0 is a dense subspace of E then the corresponding weak* topologies on $E_0' \cong E'$ do agree on bounded sets, and in typical applications this is all that matters. This observation is important because, although we are mainly interested in Banach space preduals, the weak* topology coming from a countable-dimensional predual (meaning that it is the linear span of a countable set of vectors) tends to be easier to handle, as illustrated in the following result.

Theorem 4.3. *Let E_0 be a countable-dimensional normed vector space and let E be its completion.*

(a) *The weak* topology on E_0' is metrizable and second countable, and according to this topology $[E_0']_1$ is compact.*

(b) *The weak* topologies on $E_0' \cong E'$ agree on bounded sets.*

Proof. (a) Say that E_0 is the linear span of the sequence (x_n). As we observed above, the weak topology on E_0' determined by evaluating on the x_n equals the weak topology determined by evaluating on all elements of E_0. So metrizability and second countability of E_0' follow from Theorem 1.28 and Proposition 1.34.

To verify compactness of the unit ball, without loss of generality assume $\|x_n\| \leq 1$ for all n. Then the map $\phi : f \mapsto (f(x_n))$ is an injection from $[E_0']_1$ into the infinite power $[\mathbf{F}]_1^{\mathbf{N}}$, and the weak topology on $[E_0']_1$ determined by evaluating on the x_n agrees with the restriction of the product topology on $[\mathbf{F}]_1^{\mathbf{N}}$ to the image of $[E_0']_1$. Since $[\mathbf{F}]_1^{\mathbf{N}}$ is compact we need only show that the image of $[E_0']_1$ is a closed subset. So let (f_k) be a sequence in $[E_0']_1$ and suppose $\phi(f_k) \to \mathbf{a} \in [\mathbf{F}]_1^{\mathbf{N}}$. Then $f_k(x_n) \to a_n$ for all n, and since $E_0 = \operatorname{span}\{x_n\}$ it follows that the sequence $f_k(x)$ must converge for every $x \in E_0$. Defining $f(x) = \lim f_k(x)$ for all $x \in E_0$, we then have $f \in [E_0']_1$ and $\phi(f_n) \to \phi(f)$. So $\phi([E_0']_1)$ is closed, as desired.

(b) It is enough to show that the weak* topologies on $[E_0']_1$ and $[E']_1$ agree, and it is clear that the weak* topology on $[E_0']_1$ is weaker (since fewer functions have to be continuous). So we just need to show that evaluating at any element x of E is continuous for the weak* topology on $[E_0']_1$. But we can find a sequence (x_n) in E_0 that converges to x in norm, and then $\hat{x}_n \to \hat{x}$ uniformly on $[E_0']_1$, so $\hat{x}$ must indeed be continuous on $[E_0']_1$ by Proposition 3.45. This completes the proof. $\square$

This result helps to explain how the weak* topology on the dual of any separable Banach space is well-behaved on bounded sets. For instance, Example 4.2 shows that on bounded subsets of l^∞, the weak* topology arising from the predual l^1 is the topology of pointwise convergence.

Part (b) of Theorem 4.3 immediately gives us the second assertion in part (a) for the dual of any separable Banach space. In slightly greater generality, we can observe that equality of the weak* topologies coming from E_0 and from E on bounded sets implies that the weak* topologies coming from all intermediate spaces agree on bounded sets. Since any separable normed vector space is nested between a countable-dimensional normed vector space and its completion, we can infer the second assertion in part (a) for the dual of any separable normed vector space.

Corollary 4.4. *(Banach-Alaoglu theorem) The closed unit ball of the dual of any separable normed vector space is weak* compact and metrizable.*

And thus:

Corollary 4.5. *The dual of any separable normed vector space is weak* separable.*

(Its unit ball is weak* separable by Corollary 4.4, so we can find a countable weak* dense subset of the ball of radius n for each n, and their union will be a countable weak* dense subset of the entire dual space.)

4.2 Duality

By Proposition 3.12 we know that E' is the set of linear functionals on E which are norm continuous. Conversely, we will show in this section that every weak* continuous linear functional on E' arises from evaluation on an element of E. This establishes a duality between normed vector spaces and their duals.

The only tool we need is an elementary result from linear algebra. Recall that the *codimension* of a subspace F of a vector space E is the dimension of the quotient E/F. Equivalently, it is the minimal cardinality of a set S such that $E = \mathrm{span}(F \cup S)$. The codimension of $F_1 \cap F_2$ is at most the sum of the codimensions of F_1 and F_2.

Lemma 4.6. *Let E be a vector space and let $f_1, \ldots, f_n : E \to \mathbf{F}$ be linear functionals. Suppose $f : E \to \mathbf{F}$ is a linear functional whose kernel contains the intersection of the kernels of the f_i. Then f can be expressed as a linear combination of the f_i.*

Proof. We may assume that the f_i are linearly independent, as redundant elements can be removed without affecting the intersection of their kernels. Suppose first that $\bigcap \ker f_i = \{0\}$. Since each $\ker f_i$ has codimension 1, their intersection has codimension at most n, and thus $\dim E \leq n$. But since the f_i are independent and $\dim E = \dim E'$, we must have that $\dim E' = n$ and the f_i span E'. In particular, f is a linear combination of the f_i.

Dropping the assumption that $\bigcap \ker f_i = \{0\}$, we can define $\tilde{E} = E/\bigcap \ker f_i$ and lift f and the f_i to linear functionals $\tilde{f}$ and $\tilde{f}_i$ on $\tilde{E}$. By the preceding paragraph, $\tilde{f}$ is a linear combination of the $\tilde{f}_i$, so again f is a linear combination of the f_i. $\qquad\square$

Now we show that the map $x \mapsto \hat{x}$ identifies the normed vector space E with the set of linear functionals on E' which are weak* continuous. We state this result in a slightly more general form which gives us control over weak* continuous linear functionals on subspaces of E'.

Theorem 4.7. *Let E be a normed vector space, let F be a subspace of E', and let $T : F \to \mathbf{F}$ be a linear functional whose kernel is closed in the relative weak* topology. Then there exists $x \in E$ such that $T = \hat{x}|_F$.*

Proof. If $T = 0$ then we can take $x = 0$. Otherwise, the set $C = \{f \in F : Tf = 1\}$ is a translation of $\ker T$, and since the weak* topology is translation invariant, it follows that C is closed for the relative weak* topology. Therefore its complement contains a basic relative weak* open set about the origin, i.e., there exist $x_1, \dots, x_n \in E$ and $\epsilon > 0$ such that

$$\{f \in F : |f(x_i)| < \epsilon \text{ for } 1 \leq i \leq n\} \subseteq C^c.$$

That is, $|f(x_i)| < \epsilon$ for all i implies $Tf \neq 1$.

Now let $f \in F$ and suppose $f(x_i) = 0$ for all i. Then $|af(x_i)| = 0$ for all i and all $a \in \mathbf{F}$, so that by the above $aTf = T(af) \neq 1$ for all $a \in \mathbf{F}$. This implies that $Tf = 0$. Thus, we have shown that the kernel of T contains the intersection of the kernels of the linear functionals $\hat{x}_i$ restricted to F. The lemma then implies that $T = \hat{x}|_F$ where x is some linear combination of the x_i. $\qquad\square$

This result contains a surprising amount of information about the duality between E and E'. We can harvest a number of consequences, some conceptually basic and some of more technical value.

First, we have the promised realization of E as the set of weak* continuous linear functionals on E'.

Corollary 4.8. *For any normed vector space E, the map $x \mapsto \hat{x}$ is a bijection between E and the set of weak* continuous linear functionals on E'.*

This is an immediate consequence of Theorem 4.7, taking $F = E'$ and using the fact that the kernel of any continuous linear functional must be closed.

Next, we can show that the "duality" between normed vector spaces and their duals is actually a dual equivalence in the sense of category theory. The morphisms between normed vector spaces (bounded linear maps) correspond to the morphisms between duals of normed vector spaces (weak* continuous linear maps). Given a bounded linear map $T : E \to F$ between normed vector spaces, define $T' : F' \to E'$ by $T'f = f \circ T$.

Corollary 4.9. *Let E and F be normed vector spaces and let $T : E \to F$ be a bounded linear map. Then $T' : F' \to E'$ is linear and continuous for* *the weak* topologies on F' and E', and every weak* continuous linear map from F' to E' is of this form.*

Proof. It is straightforward to check that T' is linear. To see that T' is continuous for the weak* topologies on E' and F', it will suffice to show that $\hat{x} \circ T'$ is weak* continuous from F' to $\mathbf{F}$ for each $x \in E$. But

$$(\hat{x} \circ T')(f) = \hat{x}(f \circ T) = f(Tx),$$

so $\hat{x} \circ T'$ is just evaluation on the vector $Tx \in F$, and therefore it is trivially weak* continuous.

Now let $\Phi : F' \to E'$ be any linear map which is continuous for the weak* topologies. For each $x \in E$ the map $f \mapsto (\Phi f)(x)$ is a weak* continuous linear functional on F', and hence it is given by evaluating at a unique point $Sx \in F$. We check that $S : E \to F$ is a bounded linear map. It is linear because

$$f(S(ax)) = (\Phi f)(ax) = a(\Phi f)(x) = a \cdot f(Sx) = f(a \cdot Sx)$$

and

$$f(S(x+y)) = (\Phi f)(x+y) = (\Phi f)(x) + (\Phi f)(y)$$
$$= f(Sx) + f(Sy) = f(Sx + Sy)$$

for all $f \in F'$, $a \in \mathbf{F}$, and $x, y \in E$. This implies that $S(ax) = a \cdot Sx$ and $S(x+y) = Sx + Sy$. We verify boundedness using Corollary 3.40. Thus, to check that $S([E]_1)$ is a bounded subset of F we need to show that $\sup_{x \in [E]_1} |f(Sx)|$ is finite for each $f \in F'$. But $f(Sx) = (\Phi f)(x)$, so this supremum is exactly $\|\Phi f\|$.

Finally, we have $\Phi = S'$ because $(\Phi f)(x) = f(Sx) = (S'f)(x)$ for all $f \in F'$ and $x \in E$. $\qquad\square$

We continue giving corollaries to Theorem 4.7. The next results have a slightly more technical flavor, but they still follow easily.

Corollary 4.10. *Let E be a normed vector space and let $T : E' \to \mathbf{F}$ be a linear functional whose kernel is weak* closed. Then T is weak* continuous.*

Corollary 4.11. *Let E be a normed vector space, let $F \subset E'$ be a weak* closed subspace, and let $f_0 \in E' \setminus F$. Then there exists $x \in E$ such that $f_0(x) = 1$ but $f(x) = 0$ for all $f \in F$.*

Proof. Apply Theorem 4.7 to the quotient map from $\text{span}(F \cup \{f_0\})$ to $\text{span}(F \cup \{f_0\})/F \cong \mathbf{F}$. $\qquad\square$

Corollary 4.12. *Let E be a normed vector space, let $F \subset E'$ be a weak* closed subspace, and let $f_0 \in E' \setminus F$. Then* $\operatorname{span}(F \cup \{f_0\})$ *is weak* closed.*

Proof. Let $F^\perp = \{x \in E : g(x) = 0 \text{ for all } g \in F\}$ and fix $f \in E'$. Suppose there exist $x, y \in F^\perp$ such that the vectors $(f_0(x), f(x))$ and $(f_0(y), f(y))$ in $\mathbf{R}^2$ are linearly independent; then $a = f_0(y)$, $b = -f_0(x)$ satisfy

$$f_0(ax + by) = af_0(x) + bf_0(y) = 0 \neq af(x) + bf(y) = f(ax + by),$$

and hence $\{g \in E' : g(ax + by) = 0\}$ is a weak* closed subspace that contains $\operatorname{span}(F \cup \{f_0\})$ but not f.

Otherwise, the vectors $(f_0(x), f(x))$ all lie on a line in $\mathbf{R}^2$ and hence there is a nonzero linear combination $cf_0 + df$ that satisfies $(cf_0 + df)(x) = 0$ for all $x \in F^\perp$. It then follows from Corollary 4.11 that $cf_0 + df \in F$, and therefore $f \in \operatorname{span}(F \cup \{f_0\})$. Thus, we have shown that any element of E' that is not in $\operatorname{span}(F \cup \{f_0\})$ is not in its closure, so this span must be weak* closed. $\qquad\square$

By Exercise 3.12, the closed unit ball of an infinite-dimensional Banach space cannot be norm compact. So the weak* topology is very different from the norm topology. On unbounded sets it is rather strange. For example, if E is infinite-dimensional then every weak* neighborhood of 0 in E' must contain a finite-codimensional subspace. To see this, observe that for any $x \in E$ and $\epsilon > 0$ the set $\{f \in E' : |f(x)| < \epsilon\}$ contains the codimension 1 subspace $\{f \in E' : f(x) = 0\} = \ker \hat{x}$. Since any weak* neighborhood of 0 contains the intersection of finitely many sets of this form, it contains a finite-codimensional subspace.

However, in duals of Banach spaces we usually only need to consider weak* topologies on bounded sets. One reason for this is that any weak* compact set is automatically bounded. This is an almost immediate consequence of the principle of uniform boundedness (Theorem 3.39), because the image of any weak* compact set under any point evaluation must be a compact, and hence bounded, subset of $\mathbf{F}$.

The other main reason why we can restrict attention to bounded sets is the Krein-Smulian theorem, which renders the behavior of the weak* topology on unbounded sets irrelevant for typical purposes.

Theorem 4.13. *(Krein-Smulian theorem) Let E be a separable Banach space and let $F \subseteq E'$ be a subspace of its dual. Suppose that $F \cap [E']_1$ is weak* closed. Then F is weak* closed.*

Proof. By linearity, $F \cap [E']_a$ is weak* closed for every $a > 0$. It follows that $F \cap [E']_a$ is norm closed for every a, which implies that F is norm closed.

Let $f_0 \in E' \setminus F$; we must show that f_0 is not in the weak* closure of F. Since F is norm closed, $d(f_0, F) > 0$. Without loss of generality assume $d(f_0, F) > 1$. Let $C = F - f_0$, so that every element of C has norm greater than 1. We must show that 0 is not in the weak* closure of C.

Let $C_1 = \{f \in C : \|f\| \leq 2\} = C \cap [E']_2$. The Banach-Alaoglu theorem implies that C_1 is weak* compact. Each element of C_1 has norm greater than 1 and hence is contained in the weak* open set $U_x = \{f : \Re f(x) > 1\}$ for some $x \in [E]_1$. By compactness we can find $x_1, \ldots, x_{k_1} \in [E]_1$ such that every $f \in C_1$ satisfies $\Re f(x_j) > 1$ for some $1 \leq j \leq k_1$. In other words, C_1 is contained in the weak* open set $V_1 = U_{x_1} \cup \cdots \cup U_{x_{k_1}}$.

Inductively, for $n > 1$ define

$$C_n = (C \cap [E']_{n+1}) \setminus (V_1 \cup \cdots \cup V_{n-1}).$$

C_n is weak* compact, and inductively every element has norm greater than n. Thus, just as in the case $n = 1$ we can find $x_{k_{n-1}+1}, \ldots, x_{k_n} \in [E]_{1/n}$ such that $C_n \subseteq V_n = U_{x_{k_{n-1}+1}} \cup \cdots \cup U_{x_{k_n}}$.

Now since $\|x_j\| \to 0$ as $j \to \infty$, we have a linear map $T : E' \to c_0$ defined by $(Tf)_j = f(x_j)$. The set $T(C)$ is an affine subspace of c_0 and it does not intersect $[c_0]_1$, so its norm closure does not contain the origin and therefore by Corollary 3.29 there exists $\mathbf{a} \in l^1 \cong (c_0)'$ such that $\sum a_j f(x_j) = 1$ for all $f \in C$.

Define $y = \sum a_j x_j$. (This series is absolutely summable since $\|x_j\| \leq 1$ for all j, and hence it is summable in E.) Then for all $f \in C$ we have $f(y) = \sum a_j f(x_j) = 1$. Thus C is contained in the weak* closed set $\{f \in E' : f(y) = 1\}$, which does not contain the origin. So 0 is not in the weak* closure of C, as we needed to show. $\qquad \square$

Note that in the condition "$F \cap [E']_1$ is weak* closed" we do not have to specify whether we mean for the weak* topology on E' or for the relative weak* topology on $[E']_1$. Since $[E']_1$ is weak* closed (it is the set of $f \in E'$ such that $|f(x)| \leq 1$ for all $x \in E$), the two conditions are equivalent.

Corollary 4.14. *Let E and F be separable Banach spaces and let $T : E' \to F'$ be a linear map. Then T is continuous for the weak* topologies on E' and F' iff $f_n \to 0$ weak* in E' implies $T f_n \to 0$ weak* in F', for bounded sequences $(f_n) \subset E'$.*

Proof. The forward direction is easy; use the characterization of sequential convergence given just before Proposition 1.55. For the reverse direction, suppose $f_n \to 0$ weak* implies $T f_n \to 0$ weak*, for every bounded sequence $(f_n) \subset E'$. To prove that T is weak* to weak* continuous it will suffice to show that $\hat{y} \circ T : E' \to \mathbf{F}$ is continuous for the weak* topology on E', for every $y \in F$ (Proposition 1.22).

Fix $y \in F$ and let $T_y = \hat{y} \circ T$. We know that $f_n \to 0$ weak* in E' implies $T_y f_n \to 0$, for any bounded sequence (f_n), and replacing f_n with $f_n - f$ yields that $f_n \to f$ weak* implies $T_y f_n \to T_y f$, for any bounded sequence (f_n). Since $[E']_1$ is weak* metrizable (Corollary 4.4) it follows that $\ker T_y \cap [E']_1$ is weak* closed, and then $\ker T_y$ is weak* closed by the Krein-Smulian theorem. So T_y is weak* continuous by Corollary 4.10. $\square$

4.3   Separation theorems

More advanced tools for dealing with dual spaces tend to have a geometric flavor. The key concept is convexity. A subset A of a vector space is *convex* if for any $x, y \in A$ and any $t \in (0, 1)$ we have $tx + (1 - t)y \in A$. Verifying the following simple facts is a good exercise.

Lemma 4.15. *Let A and B be convex sets in some vector space, let $a \in \mathbf{R}$, and let $b, c > 0$. Then*

(i) $A + B = \{x + y : x \in A \text{ and } y \in B\}$ *is convex;*
(ii) aA *is convex;*
(iii) $bA + cA = (b + c)A$.

Also, observe that $A + B = \bigcup_{y \in B} A + y$. This shows that if A is weak* open then $A + B$ will also be weak* open.

The basic idea of this section is that convex sets can be separated by continuous linear functionals. This is essentially a real phenomenon, but we can get complex versions of the main results by invoking Lemma 3.26.

We first prove that convex sets can be separated from the origin. Then a simple trick will enable us to separate convex sets from each other.

Lemma 4.16. *Let U be an open convex subset of $\mathbf{R}^2$ which does not contain the origin. Then there is a line through the origin that does not intersect U.*

Proof. Using Lemma 4.15 (iii) we can see that $\bigcup_{a>0} aU$ is convex. This is an open wedge in $\mathbf{R}^2$, and since it is convex and does not contain the origin, it cannot subtend an angle greater than π. So there is a line through the origin that does not intersect it. $\square$

Theorem 4.17. *Let E be a separable normed vector space and let U be a weak* open convex subset of E' that does not contain the origin. Then there exists $x \in E$ such that $\Re f(x) > 0$ for all $f \in U$.*

Proof. We assume real scalars. For the complex case, regard E as a real vector space, identify its real and complex duals by Lemma 3.26, and apply the real version of the theorem.

By Corollary 4.5 E' is weak* separable, so let (f_n) be a sequence whose span is weak* dense in E'. We may assume that $f_1 \in U$ and no other f_n is a scalar multiple of f_1.

Let $U_1 = U$. Recursively construct open, convex sets U_n that do not contain the origin together with real scalars a_n for $n \geq 2$ as follows. Given U_{n-1}, identify the span of f_1 and f_n with $\mathbf{R}^2$; the intersection of U_{n-1} with this plane is then an open convex subset of $\mathbf{R}^2$ that does not contain the origin, though it does contain the vector f_1. By Lemma 4.16 we can find a real scalar a_n such that U_{n-1} does not intersect the line $L_n = \mathbf{R} \cdot g_n$ spanned by the vector $g_n = a_n f_1 + f_n$. Let $U_n = U_{n-1} + L_n$. Since U_n is the sum of two convex sets it is convex, and since one of the summands is open it is open. By construction it does not intersect the origin.

After this construction is completed, define F to be the weak* closure of the span of the vectors g_n for $n \geq 2$. Observe that $U_n = U + \operatorname{span}\{g_2, \ldots, g_n\}$; since U_n does not contain the origin for any n, it follows that F does not intersect U. We claim that F has codimension 1. That is because the span of $F \cup \{f_1\}$ is a weak* closed (by Corollary 4.12) subspace that contains every f_n and hence equals E'. Thus by Theorem 4.7 there exists $x \in E$ such that $\ker \hat{x} = F$. It follows that $\hat{x}(U)$ is a convex subset of $\mathbf{R}$ that does not contain the origin; possibly replacing x with $-x$, we conclude that $f(x) > 0$ for all $f \in U$. $\square$

We can easily infer a version of this result for closed convex sets, and this leads to the most useful separation result, which applies to pairs of compact convex sets.

Corollary 4.18. *Let E be a separable normed vector space and let C be a weak* closed convex subset of E' that does not contain the origin. Then there exists $x \in E$ such that $\Re f(x) > 0$ for all $f \in C$.*

Proof. Since C is closed there is a basic open neighborhood U of the origin that is disjoint from C. But basic open sets are convex (see the comment just after Definition 4.1), so $C - U$ is a weak* open convex subset of E' that contains C but does not contain the origin. Apply Theorem 4.17 to this set. $\qquad\square$

Corollary 4.19. *Let E be a separable normed vector space and let K_1 and K_2 be disjoint weak* compact convex subsets of E'. Then there exist $x \in E$ and $a, b \in \mathbf{R}$ such that*

$$\Re f(x) \le a < b \le \Re g(x)$$

for all $f \in K_1$ and $g \in K_2$.

Proof. Let $C = K_2 - K_1$. Then C is the image of the compact convex set $K_1 \times K_2$ in $E' \times E'$ under the continuous linear map $(f, g) \mapsto f - g$, so it is compact and convex. It does not contain the origin since K_1 and K_2 are disjoint.

Apply Corollary 4.18 to C to find $x \in E$ such that $\Re(g(x) - f(x)) > 0$ for all $f \in K_1$ and $g \in K_2$. Thus $\Re f(x) < \Re g(x)$ for all $f \in K_1$ and $g \in K_2$. So letting $a = \sup\{\Re f(x) : f \in K_1\}$ and $b = \inf\{\Re g(x) : g \in K_2\}$, we obtain $\Re f(x) \le a \le b \le \Re g(x)$ for all $f \in K_1$ and $g \in K_2$. But by compactness, the supremum and infimum are achieved by some $f \in K_1$ and $g \in K_2$, so we must have $a < b$. $\qquad\square$

Corollary 4.19 is the main result of this section. It has many applications. For example, the following result is quite useful.

Theorem 4.20. *Let E be a separable normed vector space and let K be a weak* compact, convex subset of E'. Suppose that*

$$\|x\| \le \sup_{f \in K} \Re f(x)$$

for all $x \in E$. Then $[E']_1 \subseteq K$.

Proof. Let $g \in E' \setminus K$. According to Corollary 4.19 there exist $x \in E$ and real scalars $a < b$ such that $\Re f(x) \le a < b \le \Re g(x)$ for all $f \in K$. Taking the supremum over f then yields $\|x\| \le a$, so we must have $\|g\| > 1$. We conclude that $[E']_1 \subseteq K$. $\qquad\square$

As an example of how this result can be used, we deduce a simple criterion which tells us that a Banach space is a dual space.

Corollary 4.21. *Let E be a Banach space and let (f_n) be a separating sequence in E'. Suppose that $[E]_1$ is compact for the weak topology determined by the f_n. Then E is isometrically isomorphic to the dual of* span$\{f_n\} \subseteq E'$.

Proof. Let F be the span of the f_n. Then the map $\phi : x \to \hat{x}|_F$ is a nonexpansive linear map from E into F', and it is injective since the f_n are separating. The proof will be complete if we can show that $\phi([E]_1) = [F']_1$. But $\phi([E]_1)$ is weak* compact by hypothesis, and

$$\sup_{x \in [E]_1} \Re \hat{x}(f) = \sup_{x \in [E]_1} \Re f(x) = \|f\|$$

for all $f \in F$. So $\phi([E]_1) = [F']_1$ by Theorem 4.20. $\Box$

To illustrate Corollary 4.21, let E and F be separable normed vector spaces; we will show that $\mathcal{B}(E, F')$ is a dual space. Without loss of generality we can assume that E and F are both countable-dimensional, since replacing them by dense subspaces will not affect the Banach space $\mathcal{B}(E, F')$. Let (x_m) and (y_n) respectively be spanning sequences in E and F with $\|x_m\| = \|y_n\| = 1$ for all m and n, and for each m and n define $x_m \otimes y_n \in \mathcal{B}(E, F')'$ by

$$(x_m \otimes y_n)(T) = (Tx_m)(y_n).$$

It is straightforward to verify that $x_m \otimes y_n$ is a bounded linear functional on $\mathcal{B}(E, F')$ whose norm is at most $\|x_m\|\|y_n\| = 1$. Thus we have a countable family of bounded linear functionals, and they are separating because $(Tx_m)(y_n) = 0$ for all m and n implies $(Tx)(y) = 0$ for all $x \in E$ and $y \in F$, which implies $T = 0$. Finally, we need to check that the unit ball of $\mathcal{B}(E, F')$ is compact for the weak topology determined by the linear functionals $x_m \otimes y_n$. We follow the argument given in the proof of Theorem 4.3 (a). Thus we amalgamate the linear functionals $x_m \otimes y_n$ into a single map from $[\mathcal{B}(E, F')]_1$ into a countable power of $[\mathbf{F}]_1$, and we must show that its image is closed. To do this, suppose (T_k) is a sequence in $[\mathcal{B}(E, F')]_1$ such that $(T_k x_m)(y_n)$ converges for all m and n. It follows by linearity that $(T_k x)(y)$ converges for all $x \in E$ and $y \in F$; fixing x, we then get a linear functional Tx on F by setting $(Tx)(y) = \lim(T_k x)(y)$, and it is bounded with norm at most 1 because it is a pointwise limit of bounded linear functionals with norm at most 1. This defines a linear map $T : E \to F'$, which again is bounded with norm at most 1 because it is a pointwise limit of maps with this property. Thus there is an operator $T \in [\mathcal{B}(E, F')]_1$ which satisfies $(T_k x_m)(y_n) \to (Tx_m)(y_n)$ for all m and n, as desired.

4.4 The Krein-Milman theorem

The Krein-Milman theorem gives geometric information about weak* compact convex sets. One of the main applications is to unit balls of dual spaces. Unit balls often have "flat spots" and "corners", and analyzing this kind of structure can be helpful.

Definition 4.22. Let E be a vector space and let $K \subseteq E$ be a convex subset. An element $x \in K$ is an *extreme point* of K if it is not a convex combination of any two other elements of K. That is, for all $y, z \in K$ and $t \in (0, 1)$

$$x = ty + (1 - t)z \quad \Rightarrow \quad x = y = z.$$

We denote the set of extreme points of K by $\text{ext}(K)$.

More generally, a convex subset $A \subseteq K$ is a *face* of K if for every element x of A, any line segment in K containing x in its interior must be contained in A. That is, for all $y, z \in K$ and $t \in (0, 1)$

$$ty + (1 - t)z \in A \quad \Rightarrow \quad y, z \in A.$$

Thus, an extreme point is an element which is a face.

Observe that if K is convex and $x = ty + (1 - t)z$ for $y, z \in K$ and $t \in (0, 1)$ then the points $y' = (1 - t)x + ty$ and $z' = tx + (1 - t)z$ also belong to K and satisfy $x = \frac{1}{2}(y' + z')$. So $x \in K$ is an extreme point if and only if it is not the midpoint between two other points in K. Another way to say this is that x is extreme if $x \pm y \in K$ only for $y = 0$.

Example 4.23. Every unit vector in l^2 is an extreme point of $[l^2]_1$. To see this, let $\mathbf{a} \in l^2$ have norm 1 and let $\mathbf{b} \in l^2$ be any nonzero vector. Then the expression

$$\|\mathbf{a} + t\mathbf{b}\|^2 = \langle \mathbf{a} + t\mathbf{b}, \mathbf{a} + t\mathbf{b} \rangle = \|\mathbf{a}\|^2 + 2t\Re\langle \mathbf{a}, \mathbf{b} \rangle + t^2\|\mathbf{b}\|^2$$

is quadratic in t and it goes to $+\infty$ as $t \to \pm\infty$. So no point of the function $f(t) = \|\mathbf{a} + t\mathbf{b}\|^2$ can be a local maximum; in particular, the point $t = 0$ is not a local maximum. Thus, for any direction $\mathbf{b}$, it will either be true that moving any distance from $\mathbf{a}$ in the $\mathbf{b}$ direction will exit the unit ball, or moving any distance from $\mathbf{a}$ in the $-\mathbf{b}$ direction will exit the unit ball.

On the other hand, if $\|\mathbf{a}\| < 1$ then we can take any nonzero vector $\mathbf{b}$, and for sufficiently small t we will have $\|\mathbf{a} \pm t\mathbf{b}\| < 1$. Then $\mathbf{a}$ is the midpoint of the line segment from $\mathbf{a} - t\mathbf{b}$ to $\mathbf{a} + t\mathbf{b}$, so $\mathbf{a}$ is not an extreme point. This shows that $\text{ext}([l^2]_1)$ is the unit sphere of l^2.

Example 4.24. Let

$$[l^\infty]_1^+ = \{\mathbf{a} \in l^\infty : 0 \le a_n \le 1 \text{ for all } n\}$$

be the positive part of the unit ball of l^∞. This is a weak* compact, convex set. We identify its extreme points. Let $\mathbf{a} \in [l^\infty]_1^+$ and suppose $0 < a_n < 1$ for some n. Then for sufficiently small t we have $\mathbf{a} \pm t \cdot \mathbf{e}^n \in [l^\infty]_1^+$, so $\mathbf{a}$ cannot be an extreme point. Conversely, suppose $a_n = 0$ or 1 for all n. If $\mathbf{a} \pm \mathbf{b} \in [l^\infty]_1^+$ then it is easy to see that each b_n must be zero. This shows that $\mathbf{a}$ is an extreme point. So $\text{ext}([l^\infty]_1^+)$ consists of those sequences every entry of which is either 0 or 1.

More generally, the weak* closed faces of $[l^\infty]_1^+$ are described as follows. Let $S \subseteq T \subseteq \mathbf{N}$ and define

$$A_{S,T} = \{\mathbf{a} \in l^\infty : 1_S \le \mathbf{a} \le 1_T\}.$$

Then $A_{S,T}$ is a weak* closed face of $[l^\infty]_1^+$, and every weak* closed face is of this form.

We just referred to the weak* topology on l^∞ without specifying which predual we had in mind. The standard predual is l^1, but since we are working with bounded sets, Theorem 4.3 (b) tells us that any dense subspace of l^1 would give rise to the same weak* topology. So we are free to use the simple description of the weak* topology on l^∞ given in Example 4.2.

Example 4.25. Let X be a locally compact Hausdorff space. We identify the extreme points of $[C_0(X)]_1$. Let $f \in [C_0(X)]_1$ and suppose $|f(x)| < 1$ for some $x \in X$. Then we can find an open set U containing x and an $\epsilon > 0$ such that $|f(y)| \le 1 - \epsilon$ for all $y \in U$; without loss of generality we can assume U has compact closure. Applying Urysohn's lemma to X^*, we can find a continuous function $g : X \to [0,1]$ such that $g(x) = 1$ and $g|_{X \setminus U} = 0$. Then $f \pm \epsilon g \in [C_0(X)]_1$, showing that f is not an extreme point. Now if X is not compact then every $f \in C_0(X)$ satisfies $|f(x)| < 1$ for some $x \in X$. So unless X is compact the unit ball of $C_0(X)$ has no extreme points.

Conversely, suppose X is compact and let $f \in [C(X)]_1$ satisfy $|f(x)| = 1$ for all $x \in X$. Let $g \in C(X)$ and assume $f \pm g \in [C(X)]_1$. Then for all $x \in X$ we have $|f(x)| = 1$ and $|f(x) \pm g(x)| \le 1$. This implies that $g(x) = 0$ for all x, and we conclude that f is an extreme point. So the extreme points of $[C(X)]_1$ are precisely those functions with unit modulus at every point.

In particular, in reference to a comment we made in Section 4.1, the Banach spaces $c_0 \cong C_0(\mathbf{N})$ and $c \cong C(\mathbf{N}^*)$ cannot be isometrically isomorphic; in fact, if X is locally compact Hausdorff but not compact and

Y is compact Hausdorff then $C_0(X)$ cannot be isometrically isomorphic to $C(Y)$. Any isometric isomorphism would carry the unit ball of one space onto the unit ball of the other and would take extreme points to extreme points. But $[C_0(X)]_1$ has no extreme points and $[C(Y)]_1$ always has the extreme point 1_Y.

We now proceed to the Krein-Milman theorem, which essentially says that a weak* compact convex set is determined by its extreme points.

Lemma 4.26. *Assume* $\mathbf{F} = \mathbf{R}$. *Let E be a normed vector space, let $K \subseteq E'$ be a weak* compact convex set, and let $x \in E$. Let $b = \sup\{f(x) : f \in K\}$. Then $K \cap \hat{x}^{-1}(b)$ is a nonempty weak* closed face of K.*

Proof. Let $f, g \in K$ and $t \in (0, 1)$. Then

$$\hat{x}(tf + (1-t)g) = tf(x) + (1-t)g(x) \le tb + (1-t)b = b,$$

with equality only if $f(x) = g(x) = b$. So $tf + (1-t)g \in K \cap \hat{x}^{-1}(b)$ implies $f, g \in K \cap \hat{x}^{-1}(b)$, which shows that $K \cap \hat{x}^{-1}(b)$ is a face of K. It is nonempty by compactness, and closed because it is the intersection of two closed sets. $\square$

A *convex combination* of finitely many points $x_1, \ldots, x_n$ in some vector space is a linear combination $\sum a_i x_i$ with each $a_i \ge 0$ and $\sum a_i = 1$. The *convex hull* of a subset A of a vector space is the set of all convex combinations of elements of A; this is the smallest convex set that contains A.

Theorem 4.27. *(Krein-Milman theorem) Let E be a separable normed vector space and let K be a weak* compact convex subset of E'. Then K is the weak* closure of the convex hull of its set of extreme points.*

Proof. As we noted near the beginning of Section 4.2, any weak* compact set must be bounded. So by Theorem 4.3 (b) we may assume E is countable-dimensional. We can assume $\mathbf{F} = \mathbf{R}$ by appealing to Lemma 3.26 to infer that when the real and complex duals of E are identified their weak* topologies match up. (A complex function is continuous if and only if its real and imaginary parts are continuous.)

Let K_0 be the weak* closure of the convex hull of $\mathrm{ext}(K)$. It is clear that $K_0 \subseteq K$; suppose $K_0 \ne K$. Let $f_0 \in K \setminus K_0$. Then by Corollary 4.19 there is a real number a and $x \in E$ such that $f(x) \le a < f_0(x)$ for all $f \in K_0$. Let $b = \sup\{f(x) : f \in K\}$; then by Lemma 4.26 the set $A = K \cap \hat{x}^{-1}(b)$ is a nonempty closed face of K which is disjoint from K_0.

Now let (U_n) be a countable base for the weak* topology, and construct a decreasing sequence of closed faces $A_0 \supseteq A_1 \supseteq \cdots$ as follows. First, let $A_0 = A$. Given A_n, if there is a nonempty closed face of K that is contained in A_n and disjoint from U_n, let A_{n+1} be such a face; otherwise, let $A_{n+1} = A_n$. Now let $B = \bigcap A_n$. It is clear that B is a closed face contained in A, and it is nonempty by compactness. We claim that it consists of a single point; since B is a face of K this point would then be an extreme point of K, yielding a contradiction.

To prove the claim, suppose $f, g \in B$ are distinct. Find $y \in E$ such that $f(y) < g(y)$ and apply Lemma 4.26. This produces a closed face B' of B which does not contain f. But then we must have $f \in U_n$ for some U_n disjoint from B'. This would force A_{n+1} to be disjoint from U_n, and hence it could not contain f, a contradiction. This verifies the claim and completes the proof. $\square$

One basic consequence of the Krein-Milman theorem is that nonempty weak* compact convex sets always have extreme points. It is easy to give examples of nonempty convex sets which have no extreme points; for instance, open convex sets never have extreme points. A simple example of a closed convex set with no extreme points is any closed half-plane in $\mathbf{R}^2$.

4.5   The Riesz-Markov theorem

In this section we will identify the duals of the Banach spaces $C(X)$ and $C_0(X)$ that we discussed in Chapter 3. The remarkable conclusion is that the duals of these spaces can be realized as spaces of scalar-valued measures. This result, known as the Riesz-Markov theorem, reveals a surprisingly fundamental connection between topology and measure theory.

In order to prove the Riesz-Markov theorem we will have to construct a measure corresponding to a given bounded linear functional on $C(X)$. This is going to require a technique that is different from the one used in Section 2.3 because in the present setting there is no natural ring of sets on which to define a premeasure in the sense of Definition 2.12. Instead, we will start with data that is initially given on the open sets for some topology.

Definition 4.28. Let X be a topological space with topology $\mathcal{T}$. A *Borel premeasure* on X is a function $\mu_0 : \mathcal{T} \to [0, \infty)$ satisfying the conditions

(i) $\mu_0(\emptyset) = 0$;

(ii) $U \subseteq V$ implies $\mu_0(U) \leq \mu_0(V)$;

(iii) $\mu_0(U) + \mu_0(V) = \mu_0(U \cup V) + \mu_0(U \cap V)$ for all U and V;

(iv) for all U and all $\epsilon > 0$ there exists V such that $U \cup V = X$ and $\mu_0(U \cap V) \leq \epsilon$.

Note that $\mu_0(X) < \infty$ according to this definition, and by monotonicity this implies that μ_0 must be bounded. The definition could be modified to allow for infinite measures, but this introduces some minor complications into the development below, and we are not so interested in infinite measures here because we intend to consider Banach spaces of measures in which the norm of a positive measure μ is given by $\|\mu\| = \mu(X)$. So infinite measures are naturally excluded.

Also observe that Borel premeasures are subadditive because

$$\mu_0(U \cup V) = \mu_0(U) + \mu_0(V) - \mu_0(U \cap V) \leq \mu_0(U) + \mu_0(V).$$

Extending Borel premeasures to measures is fairly straightforward, though verifying that the construction works is a bit tedious.

Theorem 4.29. *Suppose μ_0 is a Borel premeasure on a compact Hausdorff space X. Let Ω be the family of sets $A \subseteq X$ with the property that for every $\epsilon > 0$ there exist open sets U and V satisfying $A \subseteq U$, $A^c \subseteq V$, and $\mu_0(U \cap V) \leq \epsilon$. Then Ω is a σ-algebra that contains every open set, and*

$$\mu(A) = \inf\{\mu_0(U) : A \subseteq U\}$$

is a complete measure defined on Ω which extends μ_0. Every set in Ω is a G_δ set minus a null set.

Proof. Condition (iv) of Definition 4.28 ensures that every open set belongs to Ω, and μ agrees with μ_0 on the open sets by monotonicity of Borel premeasures.

Ω is stable under complements because the condition defining membership in Ω is symmetric under complementation. We check stability under finite unions. Let $A_1, A_2 \in \Omega$ and let $\epsilon > 0$. Find U_1, V_1, U_2, V_2 witnessing the fact that A_1 and A_2 belong to Ω. Then let $U = U_1 \cup U_2$ and $V = V_1 \cap V_2$. We have $A_1 \cup A_2 \subseteq U$ and $(A_1 \cup A_2)^c = A_1^c \cap A_2^c \subseteq V$, and

$$\mu_0(U \cap V) \leq \mu_0((U_1 \cap V_1) \cup (U_2 \cap V_2))$$
$$\leq \mu_0(U_1 \cap V_1) + \mu_0(U_2 \cap V_2) \leq 2\epsilon,$$

which is enough to verify that $A_1 \cup A_2$ is in Ω.

Next let us check that μ is finitely additive. Suppose $A_1, A_2 \in \Omega$ are disjoint. We have $\mu(A_1 \cup A_2) \leq \mu(A_1) + \mu(A_2)$ since $A_1 \subseteq U_1$, $A_2 \subseteq U_2$ implies $A_1 \cup A_2 \subseteq U_1 \cup U_2$, and hence

$$\mu(A_1 \cup A_2) \leq \mu_0(U_1 \cup U_2) \leq \mu_0(U_1) + \mu_0(U_2);$$

taking an infimum over U_1 and U_2 yields the stated inequality. For the reverse inequality, let $\epsilon > 0$ and find open sets U_1 and V_1 witnessing the fact that $A_1 \in \Omega$. Then for any open set U containing $A_1 \cup A_2$, the sets $U \cap U_1$ and $U \cap V_1$ respectively contain A_1 and A_2, so we have

$$
\begin{aligned}
\mu(A_1) + \mu(A_2) &\leq \mu_0(U \cap U_1) + \mu_0(U \cap V_1) \\
&= \mu_0\big((U \cap U_1) \cup (U \cap V_1)\big) + \mu_0\big((U \cap U_1) \cap (U \cap V_1)\big) \\
&= \mu_0(U) + \mu_0(U \cap U_1 \cap V_1) \\
&\leq \mu_0(U) + \epsilon.
\end{aligned}
$$

As U and ϵ are arbitrary, this yields $\mu(A_1) + \mu(A_2) \leq \mu(A_1 \cup A_2)$.

To complete the proof that Ω is a σ-algebra, by Proposition 2.2 it will suffice to show that Ω is stable under countable disjoint unions. So let (A_n) be a sequence of disjoint sets in Ω. We use finite additivity of μ liberally in the following argument. Fix $\epsilon > 0$ and for each n find an open set U_n containing A_n such that $\mu(U_n) \leq \mu(A_n) + \frac{\epsilon}{2^n}$. Now $U = \bigcup U_n$ is open, so there exists an open set V such that $U \cup V = X$ and $\mu(U \cap V) \leq \epsilon$. Since V^c is compact and contained in $\bigcup_{n=1}^{\infty} U_n$, it must be contained in $\bigcup_{n=1}^{N} U_n$ for some N. It follows that

$$
\mu\left(U \setminus \bigcup_{n=1}^{N} U_n\right) \leq \mu(U \cap V) \leq \epsilon;
$$

finding $\tilde{V}$ such that $(\bigcup_{n=1}^{N} A_n)^c \subseteq \tilde{V}$ and $\mu((\bigcup_{n=1}^{N} A_n)^c) \geq \mu(\tilde{V}) - \epsilon$, we then have $\tilde{V}^c \subseteq \bigcup_{n=1}^{N} A_n \subseteq U$ and

$$
\begin{aligned}
\mu(U) &= \mu\left(\bigcup_{n=1}^{N} U_n\right) + \mu\left(U \setminus \bigcup_{n=1}^{N} U_n\right) \\
&\leq \mu\left(\bigcup_{n=1}^{N} A_n\right) + 2\epsilon \leq \mu(\tilde{V}^c) + 3\epsilon.
\end{aligned}
$$

It follows that $\mu(U \cap \tilde{V}) \leq 3\epsilon$. Since ϵ was arbitrary, this shows that $\bigcup A_n$ lies in Ω. We conclude that Ω is a σ-algebra, and we also get that μ is countably additive because

$$
\mu\left(\bigcup_{n=1}^{\infty} A_n\right) \leq \mu(U) \leq \mu\left(\bigcup_{n=1}^{N} A_n\right) + 2\epsilon = \sum_{n=1}^{N} \mu(A_n) + 2\epsilon,
$$

and taking ϵ to zero yields $\mu(\bigcup A_n) \leq \sum \mu(A_n)$. The reverse inequality follows directly from finite additivity since $\sum_{n=1}^{N} \mu(A_n) = \mu(\bigcup_{n=1}^{N} A_n) \leq \mu(\bigcup_{n=1}^{\infty} A_n)$ for all N.

It is more or less immediate from the definition of μ that every set in Ω is contained in a G_δ set with the same measure. For completeness, suppose

$A \in \Omega$ and $\mu(A) = 0$, and given $\epsilon > 0$ find an open set U containing A such that $\mu(U) \leq \epsilon$. Then this U and $V = \emptyset$ verify the criterion for membership in Ω for every subset of A. □

By Theorem 2.9 (d), saying that a measurable set A is a G_δ set minus a null set is equivalent to saying that for every $\epsilon > 0$ there is an open set U containing A such that $\mu(U \setminus A) \leq \epsilon$. Observe that if X is compact then applying the same argument to A^c yields a compact set K contained in A such that $\mu(A \setminus K) \leq \epsilon$. As in Theorem 2.19 (i), we say that μ is *regular*.

We now have all the tools we need to handle positive measures, but we still need to deal with the scalar-valued case. Recall that a set A is said to be null for a scalar-valued measure μ if $\mu(B) = 0$ for all measurable $B \subseteq A$. Equivalently, A is null for μ if and only if $|\mu|(A) = 0$. Similarly, we say that μ is *complete* if every subset of a null set is measurable, which is equivalent to saying that $|\mu|$ is complete. We can complete any scalar-valued measure by the same simple construction that accomplishes this for positive measures; see Proposition 2.11. If before completing every measurable set is a G_δ set minus a null set, this will still be true after completing.

Definition 4.30. Let X be a topological space. A *scalar-valued Borel measure* on X is a complete scalar-valued measure defined on a σ-algebra that contains every open set, with the property that every measurable set can be expressed as a G_δ set minus a null set.

We can define a *positive* Borel measure to be a scalar-valued Borel measure that only takes values in $[0, \infty)$. As we mentioned earlier, we will only consider finite positive Borel measures, but in order to avoid confusion we will continue to include the qualifier "finite".

Completeness is a natural and desirable condition. However, imposing this condition slightly complicates the vector space structure of the set of scalar-valued measures, because taking the sum of two complete measures in the obvious way produces a measure which need not be complete. (For instance, consider sums of the form $\mu + (-\mu)$.) Fortunately, this is not a serious problem, because we can simply complete the sum. There is no danger of any inconsistency in this prescription because of the following result.

Proposition 4.31. *Every scalar-valued Borel measure is determined by its values on open sets.*

Proof. Let μ and μ' be scalar-valued Borel measures on some topological space and suppose they agree on open sets. By Theorem 2.9 (d) and the fact that any scalar-valued measure is a linear combination of positive measures, it follows that μ and μ' agree on all G_δ sets. Since every measurable set for any Borel measure can be written as a G_δ set minus a null set, it follows that every null set is contained in a null G_δ set. Thus the fact that μ and μ' agree on the G_δ sets entails that they have the same null sets. A second appeal to the fact that every set is a G_δ set minus a null set now shows that $\mu = \mu'$. $\qquad\qquad\square$

We can now consider the space of all Borel measures.

Definition 4.32. Let X be a topological space. We denote by $M(X)$ the set of all scalar-valued Borel measures on X. The sum of two measures in $M(X)$ is defined by summing pointwise on the intersection of the σ-algebras on which they are defined and then completing. We define the norm of any scalar-valued Borel measure by $\|\mu\| = |\mu|(X)$.

$M(X)$ is a Banach space. This can be checked directly, but there is no need for us to do so because we are about to prove something stronger, at least in the case of interest to us here, when X is locally compact and second countable. We will show that in this case $M(X)$ is isometrically isomorphic to the dual of $C_0(X)$.

We start by showing how elements of $M(X)$ correspond to bounded linear functionals on $C_0(X)$.

Lemma 4.33. *Let X be a second countable locally compact Hausdorff space and let μ be a scalar-valued Borel measure on X. Then the map $\phi_\mu : f \mapsto \int f \, d\mu$ is a bounded linear functional on $C_0(X)$ and its norm equals $\|\mu\|$.*

Proof. First suppose X is compact. Since $|\mu|$ is finite, every bounded measurable function is integrable, and hence every function in $C(X)$ is integrable. So $\phi_\mu(f) = \int f \, d\mu$ is well-defined. This is a linear functional by Theorem 2.30, and it is bounded because

$$\left| \int f \, d\mu \right| = \left| \int f \cdot \frac{d\mu}{d|\mu|} \, d|\mu| \right| \le \int |f| \, d|\mu| \le \|f\|_\infty |\mu|(X)$$

(using Proposition 2.47, Theorem 2.30 (iii), and Corollary 2.46). Indeed, this shows that $\|\phi_\mu\| \le \|\mu\|$. For the reverse inequality, let $A_1, \ldots, A_n$ be a measurable partition of X and let $\epsilon > 0$. (Recall that $|\mu|(X)$ can be approximated using finite measurable partitions; see the comment following Proposition 2.41.) Use regularity to find compact sets $K_i \subseteq A_i$ such that

$|\mu|(A_i \setminus K_i) \leq \frac{\epsilon}{n}$ $(1 \leq i \leq n)$. Then apply the Tietze extension theorem to find $f \in C(X)$ such that $\|f\|_\infty \leq 1$ and $f|_{K_i} = \frac{|\mu(K_i)|}{\mu(K_i)}$ for all i, using our usual convention that $\frac{0}{0} = 0$. We then have

$$\int_{\bigcup K_i} f \, d\mu = \sum |\mu(K_i)| \geq \sum |\mu(A_i)| - \epsilon$$

and

$$\left| \int_{(\bigcup K_i)^c} f \, d\mu \right| \leq \int_{(\bigcup K_i)^c} |f| \, d|\mu| \leq \epsilon.$$

Since ϵ was arbitrary, it follows that $\|\phi_\mu\| \geq \sum |\mu(A_i)|$. Taking the supremum over all finite measurable partitions of X then yields $\|\phi_\mu\| \geq \|\mu\|$.

In the noncompact case, we get everything but the final inequality $\|\phi_\mu\| \geq \|\mu\|$ by working in $C(X^*)$ and extending μ to X^* by setting $\mu(\{\infty\}) = 0$. To prove this last inequality, recall that X being second countable implies that X^* is metrizable. Thus, by fixing a metric on X^* and considering the sets $\mathrm{ball}_{1/n}(\infty)^c$, we can reduce to the compact case because $|\mu|(\mathrm{ball}_{1/n}(\infty)^c) \to |\mu|(X)$ and we can extend continuous functions on $\mathrm{ball}_{1/n}(\infty)^c$ to continuous functions on X which vanish at infinity using the Tietze extension theorem. $\quad\square$

The key problem we still face is showing that every bounded linear functional on $C(X)$ arises from integrating against a Borel measure. First we reduce to the case of positive linear functionals. A linear functional $\phi : C(X) \to \mathbf{F}$ is *positive* if $f \geq 0$ implies $\phi(f) \geq 0$.

Lemma 4.34. *Let X be a compact metrizable space and suppose $\mathbf{F} = \mathbf{R}$. Then any bounded linear functional on $C(X)$ is a difference of two positive linear functionals.*

Proof. Let $K \subset C(X)'$ be the set of positive linear functionals ϕ which satisfy $\phi(1_X) \leq 1$. It is easy to see that K is bounded, convex, and weak* closed, hence weak* compact. We apply Theorem 4.20 to the set $K - K$. This set is the image of $K \times K \subset C(X)' \times C(X)'$ under the continuous linear map $(\phi, \psi) \mapsto \phi - \psi$, so it is convex and weak* compact.

Now for any $f \in C(X)$, there exists $x \in X$ such that $f(x) = \pm\|f\|_\infty$. Then the linear functional $\phi : g \mapsto \pm g(x)$ belongs to $K - K$ and satisfies $\phi(f) = \|f\|_\infty$. So Theorem 4.20 implies that $[C(X)']_1 \subseteq K - K$. By scaling, we conclude that every bounded linear functional is a difference of two positive linear functionals. $\quad\square$

We can now prove the Riesz-Markov theorem.

Theorem 4.35. *(Riesz-Markov theorem) Let X be a second countable locally compact Hausdorff space. Then every bounded linear functional on $C_0(X)$ is given by integrating against a scalar-valued Borel measure on X, and this pairing implements an isometric isomorphism between $C_0(X)'$ and $M(X)$.*

Proof.   By Lemma 4.33, integration against any Borel measure μ defines a bounded linear functional on $C_0(X)$ whose norm equals $\|\mu\|$. To complete the proof we need to show that every bounded linear functional on $C_0(X)$ arises from integration against a Borel measure. We can reduce to the compact metrizable case by embedding $C_0(X)$ in $C(X^*)$ and extending $\phi \in C_0(X)'$ to $C(X^*)$ by setting $\phi(1_{X^*}) = 0$. This reduction works because any Borel measure on X^* restricts to a Borel measure on X in an obvious way, and the integral of a function which vanishes at infinity will not change. We can also reduce to the case of real scalars by restricting the real and imaginary parts of a given bounded linear functional to the real part of $C(X)$, finding signed measures μ_1 and μ_2 which implement them, and then setting $\mu = \mu_1 + i\mu_2$.

Thus, assume real scalars, suppose X is a compact metrizable space, and let $\phi \in C(X)'$. By Lemma 4.34 we may suppose ϕ is positive. Now we have to construct a Borel measure μ and show that $\int f\, d\mu = \phi(f)$ for all $f \in C(X)$.

We construct μ using Theorem 4.29. Write $f \prec U$ if $0 \leq f \leq 1_K$ for some compact subset $K \subseteq U$. We define a Borel premeasure μ_0 by setting

$$\mu_0(U) = \sup\{\phi(f) : f \prec U\}.$$

Conditions (i) and (ii) of Definition 4.28 are easy. For part (iv), let U be an open set and let $\epsilon > 0$. Find $f \in C(X)$ such that $f \prec U$ and $\phi(f) \geq \mu_0(U) - \epsilon$. Let K be a compact subset of U with $f \leq 1_K$; then we claim that $V = K^c$ verifies condition (iv). This is because any function supported on $U \cap V$ will be disjointly supported from f, so that $g \prec U \cap V$ implies $f + g \prec U$ and hence

$$\phi(f + g) \leq \mu_0(U) \leq \phi(f) + \epsilon.$$

Thus $\phi(g) \leq \epsilon$, and we conclude that $\mu_0(U \cap V) \leq \epsilon$.

To verify condition (iii), let U and V be open sets. Given $f \prec U$ and $g \prec V$, let $K \subseteq U$ and $K' \subseteq V$ be compact sets on which f and g are supported and use Urysohn to find h and k satisfying $1_{K \cup K'} \leq h \prec$

$U \cup V$ and $1_{K \cap K'} \le k \prec U \cap V$. Then $f + g \le h + k$, and we infer that $\mu_0(U) + \mu_0(V) \le \mu_0(U \cup V) + \mu_0(U \cap V)$. For the reverse inequality, suppose $h \prec U \cup V$ and $k \prec U \cap V$. Using the fact that X is metrizable, find a sequence (U_n) of open sets such that $\overline{U}_n \subseteq U_{n+1}$ for all n and $U = \bigcup U_n$, and let (V_n) be a sequence of open sets with the same properties relative to V. Then $U \cup V = \bigcup (U_n \cup V_n)$, so any compact subset of $U \cup V$ must be contained in some $U_n \cup V_n$; thus $h \le 1_{U_n \cup V_n}$ for some n. Similarly, $k \le 1_{U_n \cap V_n}$ for some n, and we may assume that these two values of n are the same. Now find f and g satisfying $1_{U_n} \le f \prec U$ and $1_{V_n} \le g \prec V$; we then have $h + k \le f + g$, and we can conclude from this that $\mu_0(U \cup V) + \mu_0(U \cap V) \le \mu_0(U) + \mu_0(V)$. This completes the proof that μ_0 is a Borel premeasure.

Let μ be the Borel measure generated from μ_0. We must show that $\int f \, d\mu = \phi(f)$ for all $f \in C(X)$. By linearity it will suffice to consider the case that $0 \le f \le 1$. Fix $n \in \mathbf{N}$ and for $1 \le i \le n$ define

$$f_i = \min\left(\max\left(f - \frac{i-1}{n}, 0\right), \frac{1}{n}\right).$$

Thus, f_i is obtained by truncating f above at $\frac{i}{n}$ and below at $\frac{i-1}{n}$, and shifting the result down by $\frac{i-1}{n}$. Also define $U_i = f^{-1}((\frac{i}{n}, 1])$ for $0 \le i \le n$. Then $f = f_1 + \cdots + f_n$ and $\frac{1}{n} 1_{U_{i-1}} \ge f_i \ge \frac{1}{n} 1_{U_i}$ for each i. Thus $\mu(U_i) \le n \cdot \phi(f_i)$ for each i, from the definition of $\mu_0(U_i)$, and also $n \cdot \phi(f_i) \le \mu(U_{i-1})$ since $n \cdot f_i$ can be uniformly approximated by functions satisfying $f \prec U_{i-1}$ (e.g., the functions $\max(nf_i - \epsilon, 0)$ for $\epsilon > 0$). Summing over i yields

$$\frac{1}{n} \sum_{i=1}^{n} \mu(U_i) \le \phi(f) \le \frac{1}{n} \sum_{i=0}^{n-1} \mu(U_i).$$

But we also have $\frac{1}{n}\mu(U_i) \le \int f_i \, d\mu \le \frac{1}{n}\mu(U_{i-1})$ for all i, and hence

$$\frac{1}{n} \sum_{i=1}^{n} \mu(U_i) \le \int f \, d\mu \le \frac{1}{n} \sum_{i=0}^{n-1} \mu(U_i).$$

We conclude that $|\phi(f) - \int f \, d\mu| \le \frac{1}{n}\mu(U_0) - \frac{1}{n}\mu(U_n) \le \frac{1}{n}\mu(X)$; since n was arbitrary, this yields that $\int f \, d\mu = \phi(f)$. $\square$

4.6 L^1 and L^∞ spaces

In the remainder of this chapter we will investigate L^p spaces. This will provide us with a large stable of examples of dual spaces. We start in this section with the cases $p = 1$ and $p = \infty$.

Let (X, μ) be a measure space. For $1 \leq p \leq \infty$ the Banach space $L^p(X, \mu)$ will be a space of measurable functions from X into $\mathbf{F}$, modulo the identification of functions which agree almost everywhere. This identification introduces a subtlety into the definition of the norm on $L^\infty(X, \mu)$, since the straight supremum $\sup_{x \in X} |f(x)|$ is sensitive to the behavior of f on sets of measure zero. We want the "essential" supremum of f, which can be defined in a variety of ways.

Definition 4.36. Let (X, μ) be a measure space and let $f : X \to \mathbf{F}$ be measurable. The *essential range* of f, $\mathrm{ess\,ran}(f)$, is the set of $z \in \mathbf{F}$ such that $\mu(f^{-1}(U)) > 0$ for every open neighborhood U of z. That is, positive measure subsets of X are mapped into arbitrarily small regions near z. The *essential supremum* of f is the quantity

$$\|f\|_\infty = \sup\{|z| : z \in \mathrm{ess\,ran}(f)\}$$

and f is *essentially bounded* if $\|f\|_\infty < \infty$.

The next result gives two other equivalent definitions of $\|f\|_\infty$. The first is in terms of not having any positive measure on which $|f|$ is greater than $\|f\|_\infty$, and the second is in terms of stability under alterations of f on null sets.

Proposition 4.37. *Let (X, μ) be a measure space and let $f : X \to \mathbf{F}$ be measurable. Then the essential range of f is a closed subset of $\mathbf{F}$, and we have*

$$\|f\|_\infty = \sup\left\{a \in [0, \infty) : \{x \in X : |f(x)| \geq a\} \text{ has positive measure}\right\}$$
$$= \inf\left\{\sup_{x \in X} |g(x)| : g = f \text{ almost everywhere}\right\}.$$

Proof. Suppose $z \in \mathbf{F}$ is not in the essential range of f. Then there is an open neighborhood U of z such that $f^{-1}(U)$ has measure zero in X. Since U is open it is a neighborhood of each of its elements, and so no element of U can be in the essential range of f. This shows that the complement of the essential range is open, i.e., the essential range is closed.

Let $r = \sup\{a \in [0, \infty) : \{x \in X : |f(x)| \geq a\}$ has positive measure$\}$ and let $s = \inf\{\sup_{x \in X} |g(x)| : g = f$ almost everywhere$\}$. We will show that $\|f\|_\infty \leq r \leq s \leq \|f\|_\infty$.

$\|f\|_\infty \leq r$: Let $\epsilon > 0$ and fix z in the essential range of f. Then $f^{-1}(\mathrm{ball}_\epsilon(z))$ has positive measure in X, and it follows that $a = |z| - \epsilon$ belongs to the supremum that defines r. Thus $r \geq |z| - \epsilon$. Taking $\epsilon \to 0$ shows that $r \geq |z|$ for all z in the essential range of f, and hence $r \geq \|f\|_\infty$.

$r \leq s$: Let g be any function which agrees with f almost everywhere and fix $a \in [0, \infty)$ such that $\{x \in X : |f(x)| \geq a\}$ has positive measure. It follows that $\{x \in X : |g(x)| \geq a\}$ has positive measure, so certainly $\sup_{x \in X} |g(x)| \geq a$. Taking the infimum over g and the supremum over a yields $s \geq r$.

$s \leq \|f\|_\infty$: We may assume $\|f\|_\infty < \infty$. For each $z \in \mathbf{F}$ with $|z| > \|f\|_\infty$ we can find an open neighborhood U_z of z such that $f^{-1}(U_z)$ has measure zero in X. Since $\{z : |z| > \|f\|_\infty\}$ is a countable union of compact sets, we can find a countable family of open sets U_{z_i} such that $f^{-1}(U_{z_i})$ has measure zero in X for all i and $\bigcup U_{z_i}$ contains every $z \in \mathbf{F}$ with $|z| > \|f\|_\infty$. It follows that $S = f^{-1}(\bigcup U_{z_i})$ has measure zero, i.e., $|f(x)| \leq \|f\|_\infty$ almost everywhere. Defining $g(x) = f(x)$ for $x \notin S$ and $g|_S = 0$, we have $g = f$ almost everywhere and $\sup_{x \in X} |g(x)| \leq \|f\|_\infty$. Thus $s \leq \|f\|_\infty$. $\qquad \square$

Now we can define $L^1(X, \mu)$ and $L^\infty(X, \mu)$.

Definition 4.38. Let (X, μ) be a measure space.

(a) $L^1(X, \mu)$ is the set of integrable functions $f : X \to \mathbf{F}$, identifying functions which agree almost everywhere, equipped with the norm $\|f\|_1 = \int |f| \, d\mu$.

(b) $L^\infty(X, \mu)$ is the set of essentially bounded measurable functions $f : X \to \mathbf{F}$, identifying functions which agree almost everywhere, equipped with the essential sup norm $\|f\|_\infty$.

It is easy to verify that $L^1(X, \mu)$ and $L^\infty(X, \mu)$ are vector spaces and $\| \cdot \|_1$ and $\| \cdot \|_\infty$ are indeed norms (exercise).

Example 4.39. We have $L^\infty(\mathbf{N}, \mu_{\text{count}}) \cong l^\infty$ and $L^1(\mathbf{N}, \mu_{\text{count}}) \cong l^1$.

Proposition 4.40. *Let (X, μ) be a measure space. Then $L^1(X, \mu)$ and $L^\infty(X, \mu)$ are Banach spaces.*

Proof. We must show that $L^1(X, \mu)$ and $L^\infty(X, \mu)$ are complete. Recall from Lemma 3.10 that this is equivalent to the condition that every absolutely summable series is summable. Thus let (f_n) be an absolutely summable series in $L^1(X, \mu)$, i.e., suppose $\sum \|f_n\|_1 < \infty$. Define $g : X \to [0, \infty]$ by $g = \sum |f_n|$. Then $\sum_{n=1}^N |f_n| \nearrow g$ pointwise as $N \to \infty$, so

$$\sum_{n=1}^N \|f_n\|_1 = \int \sum_{n=1}^N |f_n| \to \int g$$

as $N \to \infty$ by MCT. Thus $\|g\|_1 = \sum \|f_n\|_1 < \infty$, so g is integrable. This implies that $g(x) < \infty$ for almost every $x \in X$, which means that the series $\sum f_n(x)$ converges absolutely for almost every x. So we can define $f = \sum f_n$ almost everywhere. Since $|f| \leq g$ almost everywhere we have $f \in L^1(X, \mu)$. Finally, $\|f - \sum_{n=1}^{N} f_n\|_1 \to 0$ by DCT since $|f - \sum_{n=1}^{N} f_n|$ converges to 0 pointwise almost everywhere and is dominated by g. Thus we have shown that $\sum f_n$ converges to f in $L^1(X, \mu)$, so $L^1(X, \mu)$ is complete.

Now let (f_n) be an absolutely summable series in $L^\infty(X, \mu)$, i.e., suppose that $M = \sum \|f_n\|_\infty < \infty$. Since $|f_n(x)| \leq \|f_n\|_\infty$ almost everywhere for each n, it follows that $\sum |f_n(x)| \leq M$ for almost every x. Thus the series $\sum f_n(x)$ converges absolutely for almost every x, and so we can define $f = \sum f_n$ almost everywhere. It is clear that $\|f\|_\infty \leq M < \infty$, so $f \in L^\infty(X, \mu)$, and

$$\left\| f - \sum_{n=1}^{N} f_n \right\|_\infty = \left\| \sum_{n=N+1}^{\infty} f_n \right\|_\infty \leq \sum_{n=N+1}^{\infty} \|f_n\|_\infty \to 0$$

as $N \to \infty$, so $\sum f_n$ converges to f in $L^\infty(X, \mu)$. This implies that $L^\infty(X, \mu)$ is complete. $\square$

L^∞ spaces are separable only if they are finite-dimensional. Otherwise we can find a sequence of disjoint positive measure sets A_n, and the map that takes $\mathbf{a}$ to $\sum a_n \cdot 1_{A_n}$ (interpreting the sum in the obvious way) is an isometric embedding of l^∞ into $L^\infty(X, \mu)$. Since l^∞ is nonseparable, $L^\infty(X, \mu)$ must be nonseparable too.

On the other hand, we have a simple criterion for separability of L^1 spaces.

Definition 4.41. A σ-finite measure space (X, μ) is *measurably separable* if there is a sequence of finite measure sets (A_n) with the property that for every finite measure set $A \subseteq X$ and every $\epsilon > 0$ there exists n such that $\mu(A \triangle A_n) \leq \epsilon$. We say that the sequence (A_n) is *measurably dense* in X.

For instance, $\mathbf{R}^n$ with Lebesgue measure is measurably separable. We can use Theorem 2.19 (iii) to show that the family of finite unions of open boxes with rational coordinates is measurably dense.

Lemma 4.42. *Let (X, μ) be a measure space. Then the simple functions in $L^1(X, \mu)$ are dense in $L^1(X, \mu)$.*

Proof. Let $f \in L^1(X, \mu)$ and suppose $f \geq 0$. Then there is a sequence of positive simple functions h_n which increase pointwise to f (see the comment preceding Theorem 2.27). By MCT we have $\int h_n \to \int f$, and hence $\|f - h_n\|_1 = \int (f - h_n) \to 0$ as $n \to \infty$. Thus every positive function in $L^1(X,\mu)$ can be approximated in L^1 norm by simple functions in $L^1(X,\mu)$. Taking linear combinations yields the same result for any function in $L^1(X,\mu)$. $\square$

(We have to say "simple functions in $L^1(X,\mu)$" because if $\mu(X) = \infty$ then there are simple functions which are not integrable.)

Proposition 4.43. *Let (X,μ) be a σ-finite measure space. Then $L^1(X,\mu)$ is separable iff (X,μ) is measurably separable.*

Proof. Suppose $L^1(X,\mu)$ is separable and let (f_n) be a dense sequence of functions. For each n let $A_n = f_n^{-1}(\mathrm{ball}_{1/2}(1))$. We claim that these sets are measurably dense in X. To see this, given $A \subseteq X$ with $\mu(A) < \infty$ and $\epsilon > 0$, find n such that $\|1_A - f_n\|_1 \le \epsilon$. Then $\mu(A \triangle A_n) \le 2\epsilon$ because $|1_A - f_n| \ge \frac{1}{2}$ on $A \triangle A_n$ and therefore

$$\epsilon \ge \int_{A \triangle A_n} |1_A - f_n| \ge \frac{1}{2}\mu(A \triangle A_n).$$

This proves the claim and shows that (X,μ) is measurably separable.

Conversely, suppose (X,μ) is measurably separable and let (A_n) be a measurably dense sequence. We claim that the linear span of the functions 1_{A_n} is dense in $L^1(X,\mu)$; by Lemma 3.11 this is enough. But if $A \subseteq X$ is any finite measure set and $\epsilon > 0$, we can find n such that

$$\|1_A - 1_{A_n}\|_1 = \mu(A \triangle A_n) \le \epsilon.$$

This shows that the closed linear span of the functions 1_{A_n} contains the characteristic function of every finite measure set; taking linear combinations, we see that it contains every simple function in $L^1(X,\mu)$, and by Lemma 4.42 it therefore contains all of $L^1(X,\mu)$, as desired. $\square$

Although $L^\infty(X,\mu)$ is typically nonseparable, if (X,μ) is σ-finite then $L^\infty(X,\mu)$ is a dual Banach space. L^1 spaces are typically not dual spaces (see Exercise 4.23).

Theorem 4.44. *Let (X,μ) be a σ-finite measure space. Then $(L^1(X,\mu))' \cong L^\infty(X,\mu)$.*

Proof. Define a map $T : L^\infty(X,\mu) \to (L^1(X,\mu))'$ by setting $(Tf)(g) = \int fg$ for $f \in L^\infty(X,\mu)$ and $g \in L^1(X,\mu)$. Observe first that the integral makes sense because $|fg| \le \|f\|_\infty |g|$, and hence fg is integrable. Next, observe that

$$\left| \int fg \right| \le \int |fg| \le \|f\|_\infty \int |g| = \|f\|_\infty \|g\|_1,$$

so Tf is a bounded linear functional on $L^1(X, \mu)$ with $\|Tf\| \leq \|f\|_\infty$ for each $f \in L^\infty(X, \mu)$.

The map T is clearly linear, and the preceding shows that T is non-expansive. To see that T is an isometry, fix a nonzero $f \in L^\infty(X, \mu)$, let $0 < \epsilon < \|f\|_\infty$, and let A be a positive measure subset of X such that $|f(x)| \geq \|f\|_\infty - \epsilon$ for all $x \in A$. Since X is σ-finite, every set of infinite measure contains a subset of finite but nonzero measure, so we may assume the measure of A is finite. Thus $g = \frac{|f|}{f} \cdot 1_A$ is in $L^1(X, \mu)$ and we have

$$\|g\|_1 = \int |g| = \int 1_A = \mu(A)$$

and

$$|Tf(g)| = \left| \int fg \right| = \int_A |f| \geq (\|f\|_\infty - \epsilon)\mu(A).$$

This shows that $\|Tf\| \geq \|f\|_\infty - \epsilon$, and taking $\epsilon \to 0$ and combining the result with the previous inequality yields $\|Tf\| = \|f\|_\infty$. Thus T is an isometry from $L^\infty(X, \mu)$ into $(L^1(X, \mu))'$.

Finally, we must show that T is surjective. To do this let $\phi \in (L^1(X, \mu))'$. Suppose first that $\mu(X) < \infty$. Then for every measurable $A \subseteq X$ define $\nu(A) = \phi(1_A)$. This is a scalar-valued measure on X which is absolutely continuous with respect to μ, since $\mu(A) = 0$ implies that $1_A = 0$ almost everywhere and hence that $\phi(1_A) = 0$. Thus by the Radon-Nikodym theorem there exists $f \in L^1(X, \mu)$ such that $\nu(A) = \int_A f \, d\mu$ for all A.

We claim that $\|f\|_\infty \leq \|\phi\|$. To see this let $z \in \mathrm{ess\,ran}(f)$ and suppose $|z| > \|\phi\|$. Let $\epsilon < |z| - \|\phi\|$ and observe that $A = f^{-1}(\mathrm{ball}_\epsilon(z))$ has positive measure. But then 1_A satisfies $\|1_A\|_1 = \mu(A)$,

$$|\phi(1_A)| = \left| \int_A f \, d\mu \right| = \left| \int_A af \, d\mu \right|$$

where $a = \frac{|z|}{z}$, and $\Re af(x) \geq |z| - \epsilon$ for all $x \in A$, so that $|\phi(1_A)| \geq (|z| - \epsilon)\mu(A)$. This implies $\|\phi\| \geq |z| - \epsilon$, contradicting our choice of ϵ. We conclude that the essential range of f is contained in the ball about 0 of radius $\|\phi\|$, and in particular $f \in L^\infty(X, \mu)$. Now Tf and ϕ agree on characteristic functions, and hence they agree on simple functions by linearity and on all of $L^1(X, \mu)$ by Lemma 4.42. Thus $\phi = Tf$, and we have shown that T is surjective.

If $\mu(X) = \infty$, we can appeal to the weighting technique described just before Theorem 2.45. In the notation given there, the map $g \mapsto gw$ de-scribes an isometric isomorphism between $L^1(X, \mu)$ and $L^1(X, \tilde{\mu})$. Since

$L^\infty(X, \mu)$ literally equals $L^\infty(X, \tilde{\mu})$, we immediately infer from the finite measure case that $(L^1(X, \mu))' \cong L^\infty(X, \mu)$ (and the pairing between $L^1(X, \mu)$ and $L^\infty(X, \mu)$ is given by the same formula $\int fg \, d\mu$ because $\int f(gw) \, d\tilde{\mu} = \int fg \, d\mu$). □

Since $L^\infty(X, \mu)$ is typically nonseparable, its structure as a Banach space is somewhat obscure. However, the existence of a weak* topology is a powerful tool. The following corollary illustrates this comment.

A *boundedly complete lattice* is a partially ordered set with the property that any nonempty subset has a least upper bound and a greatest lower bound, provided it has any upper bound and any lower bound. For example, **R** with its usual ordering is a boundedly complete lattice; **Q** is not. Assuming real scalars, we define an ordering on $L^\infty(X, \mu)$ by setting $f \le g$ if $f(x) \le g(x)$ for almost every x.

Corollary 4.45. *Let (X, μ) be a measurably separable σ-finite measure space and suppose $\mathbf{F} = \mathbf{R}$. Then $L^\infty(X, \mu)$ is a boundedly complete lattice.*

Proof. Let S be a nonempty bounded family of real-valued functions; we prove it has a least upper bound. Without loss of generality we may assume that $f, g \in S$ implies $\max(f, g) \in S$. For each $f \in S$ let C_f be the weak* closure of $\{g \in S : g \ge f\}$. Then each C_f is weak* compact, and the intersection of C_f and C_g contains $C_{\max(f,g)}$. It follows that $\bigcap_{f \in S} C_f$ is nonempty. Let $F \in \bigcap C_f$.

We claim that F is an upper bound for S. Suppose not; then for some $f \in S$ it is not true that $f \le F$ almost everywhere. Then there must exist $\epsilon > 0$ and a positive measure set $A \subseteq X$ such that $f \ge F + \epsilon$ on A. We may assume $\mu(A) < \infty$. Then for every $g \in C_f$ we have $g \ge F + \epsilon$ on A and hence

$$\int g \cdot 1_A \ge \int (F + \epsilon) \cdot 1_A = \int F \cdot 1_A + \epsilon\mu(A).$$

But this contradicts the fact that F is in the weak* closure of C_f. This proves the claim.

Finally, we verify that F is the least upper bound of S. Let $F' \in L^\infty(X, \mu)$ satisfy $f \le F'$ a.e. for all $f \in S$ and suppose $F \le F'$ a.e. fails. Find $\delta > 0$ and a finite positive measure set B such that $F \ge F' + \delta$ on B. It follows that $F \ge f + \delta$ on B for all $f \in S$, so that

$$\int F \cdot 1_B \ge \int (f + \delta) \cdot 1_B = \int f \cdot 1_B + \delta\mu(B)$$

for all $f \in S$. This contradicts the fact that F is in the weak* closure of S. We conclude that F is a least upper bound for S. The existence of greatest lower bounds is proven similarly. $\square$

Another way having a weak* topology makes $L^\infty(X, \mu)$ more manageable is by enabling us to approximate arbitrary bounded measurable functions with functions belonging to some nice class. For instance, under reasonable hypotheses we can approximate every element of $L^\infty(X, \mu)$ by continuous functions.

Proposition 4.46. *Let X be a locally compact metrizable space and let μ be a finite positive Borel measure on X. Then $[C_0(X)]_1$ is weak* dense in $[L^\infty(X, \mu)]_1$.*

Proof. First, observe that (X, μ) is measurably separable. A measurably dense sequence can be constructed by taking finite unions of sets belonging to a countable base for X. Therefore $L^1(X, \mu)$ is separable (Proposition 4.43), so we can prove the proposition by showing that $\|f\|_1 = \sup_{g \in [C_0(X)]_1} \Re \int fg$ for all $f \in L^1(X, \mu)$ and invoking Theorem 4.20.

It will be enough to show this for simple functions h since the latter are dense in $L^1(X, \mu)$ (Lemma 4.42). Indeed, we may assume that $h = \sum_{i=1}^n a_i 1_{K_i}$ where the K_i are compact and disjoint, since every simple function is approximated in L^1 norm by such functions. Given such a function h, use the Tietze extension theorem to find $g \in [C_0(X)]_1$ such that $g|_{K_i} = \frac{|a_i|}{a_i}$ for each i. We then have

$$\|h\|_1 = \sum |a_i| \mu(K_i) = \int hg,$$

which suffices to prove the proposition. $\square$

4.7 L^p spaces

In this section we present the basic theory of L^p spaces for p intermediate between 1 and ∞. By far the most important of these is the case $p = 2$, when $L^p(X, \mu)$ is a Hilbert space, a topic we will pursue further in Chapter 5.

Definition 4.47. Let (X, μ) be a measure space and let $1 < p < \infty$. Then $L^p(X, \mu)$ is the set of measurable functions $f : X \to \mathbf{F}$ such that $\int |f|^p \, d\mu <$

∞, identifying functions which agree almost everywhere, equipped with the norm $\|f\|_p = (\int |f|^p \, d\mu)^{1/p}$.

Taking (X, μ) to be $\mathbf{N}$ with counting measure, we obtain a family of sequence spaces which are usually denoted l^p. In particular, we recover l^2 when $p = 2$.

Our first order of business is to check that $\|f\|_p$ is a norm. The equality $\|af\|_p = |a| \cdot \|f\|_p$ for $a \in \mathbf{F}$ is easy. We will prove the triangle inequality using a generalization of the inequality $|\langle \mathbf{a}, \mathbf{b} \rangle| \le \|\mathbf{a}\|_2 \|\mathbf{b}\|_2$ which we encountered in Chapter 3 when introducing the inner product on l^2. This result involves the *conjugate exponent* of $p \in (1, \infty)$, which is the number $q = \frac{p}{p-1}$ that satisfies $\frac{1}{p} + \frac{1}{q} = 1$. Notice that as p increases from 1 to ∞ its conjugate exponent decreases from ∞ to 1, with equality occuring at $p = q = 2$.

Lemma 4.48. *(Hölder's inequality) Let $1 < p < \infty$, let (X, μ) be a measure space, and let $f, g : X \to \mathbf{F}$ be measurable. Then $\|fg\|_1 \le \|f\|_p \|g\|_q$, where q is the conjugate exponent to p.*

Proof. The proof relies on the inequality $st \le \frac{s^p}{p} + \frac{t^q}{q}$ for $s, t \ge 0$. This can be proven by fixing s and minimizing $\frac{t^q}{q} - st$; the minimum occurs at $t = s^{1/(q-1)}$, when $\frac{t^q}{q} - st = \frac{s^p}{q} - s^p = -\frac{s^p}{p}$. Thus $\frac{t^q}{q} - st \ge -\frac{s^p}{p}$ for all $s, t \ge 0$. Now $\|f\|_p = 0$ implies $f = 0$ almost everywhere, which makes the inequality trivial, and the same is true if $\|g\|_q = 0$, so we may assume that $\|f\|_p \ne 0 \ne \|g\|_q$. Thus let $f' = \frac{1}{\|f\|_p} f$ and $g' = \frac{1}{\|g\|_q} g$; the preceding inequality yields

$$\|f'g'\|_1 = \int |f'g'| \le \int \left(\frac{1}{p}|f'|^p + \frac{1}{q}|g'|^q \right) = \frac{1}{p} + \frac{1}{q} = 1.$$

We conclude that $\|fg\|_1 \le \|f\|_p \|g\|_q$. $\qquad\qquad\square$

Lemma 4.49. *(Minkowski's inequality) Let $1 < p < \infty$, let (X, μ) be a measure space, and let $f, g : X \to \mathbf{F}$ be measurable. Then $\|f + g\|_p \le \|f\|_p + \|g\|_p$.*

Proof. Letting q be the conjugate exponent to p and using Hölder's inequality, we have

$$\int |f + g|^p \le \int |f| \cdot |f + g|^{p-1} + \int |g| \cdot |f + g|^{p-1}$$
$$\le \|f\|_p \||f + g|^{p-1}\|_q + \|g\|_p \||f + g|^{p-1}\|_q$$

$$= (\|f\|_p + \|g\|_p) \left(\int |f + g|^p \right)^{1/q},$$

which implies $\|f + g\|_p = (\int |f + g|^p)^{1/p} \leq \|f\|_p + \|g\|_p$ since $1 - \frac{1}{q} = \frac{1}{p}$. $\square$

Now that we know $\| \cdot \|_p$ is a norm, we can show that $L^p(X, \mu)$ is a Banach space.

Proposition 4.50. *Let (X, μ) be a measure space and let $1 < p < \infty$. Then $L^p(X, \mu)$ is a Banach space.*

Proof. By Lemma 4.49, $L^p(X, \mu)$ is a vector space and $\| \cdot \|_p$ is a norm. To verify completeness, let (f_n) be an absolutely summable sequence in $L^p(X, \mu)$, i.e., suppose $\sum \|f_n\|_p < \infty$. Define $g : X \to [0, \infty]$ by $g = \sum |f_n|$. Then $(\sum_{n=1}^N |f_n|)^p \nearrow g^p$ pointwise as $N \to \infty$, so

$$\int g^p = \lim_{N \to \infty} \int \left(\sum_{n=1}^N |f_n| \right)^p = \lim_{N \to \infty} \left\| \sum_{n=1}^N |f_n| \right\|_p^p$$

$$\leq \lim_{N \to \infty} \left(\sum_{n=1}^N \|f_n\|_p \right)^p < \infty$$

by MCT. Thus $g \in L^p(X, \mu)$. This implies that $g(x) < \infty$ for almost every $x \in X$, which means that the series $\sum f_n(x)$ converges absolutely for almost every x. So we can define $f = \sum f_n$ almost everywhere. Since $|f| \leq g$ we have $f \in L^p(X, \mu)$. Finally, $\int |f - \sum_{n=1}^N f_n|^p \to 0$ by DCT since these differences converge to 0 pointwise almost everywhere and are dominated by g^p. Taking pth roots, this shows that $\sum f_n$ converges to f in $L^p(X, \mu)$, so $L^p(X, \mu)$ is complete by Lemma 3.10. $\square$

Next we prove two basic density results. We already know the first half of this proposition for $L^1(X, \mu)$.

Proposition 4.51. *Let (X, μ) be a measure space and let $1 \leq p < \infty$.*

(a) The simple functions in $L^1(X, \mu)$ are dense in $L^p(X, \mu)$.

(b) If X is a compact metrizable space and μ is a finite positive Borel measure on X then $C(X)$ is dense in $L^p(X, \mu)$.

Proof. (a) Let $f \in L^p(X, \mu)$ and suppose $f \geq 0$. Then there is a sequence of positive simple functions h_n which increase pointwise to f. Observe that each h_n is supported on a set of finite measure, otherwise it (and hence f) would not belong to $L^p(X, \mu)$, so that each h_n is a simple function in $L^1(X, \mu)$. Now $|f - h_n|^p \to 0$ pointwise and $|f - h_n|^p \leq |f|^p \in L^1(X, \mu)$, so by DCT we have $\|f - h_n\|_p = (\int |f - h_n|^p)^{1/p} \to 0$ as $n \to \infty$. Thus every positive function in $L^p(X, \mu)$ can be approximated in L^p norm by simple functions in $L^1(X, \mu)$. Taking linear combinations yields the same result for any function in $L^p(X, \mu)$.

(b) By part (a) it will suffice to show that the characteristic function of any measurable set is in the L^p-norm closure of $C(X)$. Let A be such a set. Given $\epsilon > 0$, by regularity (see the comment following Definition 4.30) we can find a compact set K contained in A and an open set U containing A such that $\mu(U \setminus K) \leq \epsilon$. Then by Urysohn's lemma we can find $f \in C(X)$ such that $0 \leq f \leq 1$, $f|_K = 1$, and $f|_{X \setminus U} = 0$. This yields

$$\|1_A - f\|_p \leq \mu(U \setminus K)^{1/p} \leq \epsilon^{1/p},$$

which is enough. $\qquad\qquad\qquad\qquad\qquad\qquad\qquad\qquad\qquad\qquad\qquad\qquad\qquad\square$

The following proposition is also quite useful.

Proposition 4.52. *Let (X, μ) be a measure space and let $1 \leq p < \infty$. Suppose $f_n \to f$ in $L^p(X, \mu)$. Then some subsequence of (f_n) converges to f almost everywhere.*

Proof. We will show that $f_n \to f$ a.e. under the assumption that $\|f - f_n\|_p \leq 4^{-n}$ for all n. The result follows because any convergent sequence has subsequences which converge arbitrarily fast.

Under this assumption, the measure of the set $A_n = \{x : |f(x) - f_n(x)| \geq 2^{-n}\}$ is at most 2^{-n} since

$$4^{-n} \geq \|f - f_n\|_p \geq \left(\int_{A_n} |f - f_n|^p\right)^{1/p} \geq 2^{-n}(\mu(A_n))^{1/p}$$

and hence $\mu(A_n) \leq (2^{-n})^p \leq 2^{-n}$. Now for any $N \in \mathbf{N}$, if $x \notin \bigcup_{n=N+1}^{\infty} A_n$ then $|f(x) - f_n(x)| < 2^{-n}$ for all $n > N$, and hence $f_n(x) \to f(x)$. But $\mu(\bigcup_{n=N+1}^{\infty} A_n) \leq 2^{-N}$, so taking $N \to \infty$, we see that $f_n \to f$ almost everywhere. $\qquad\qquad\qquad\qquad\qquad\qquad\qquad\qquad\qquad\qquad\qquad\square$

(We stated Proposition 4.52 for $p < \infty$ because when $p = \infty$ it is trivial that $f_n \to f$ in $L^\infty(X, \mu)$ implies $f_n \to f$ uniformly almost everywhere.)

We conclude with a discussion of duality for L^p spaces.

Lemma 4.53. *Let (X, μ) be a finite measure space, let $1 < p, q < \infty$ be conjugate exponents, let $f \in L^1(X, \mu)$, let $M > 0$, and suppose $|\int fg| \leq M\|g\|_q$ for all simple functions g. Then $\|f\|_p \leq M$.*

Proof.   For $n \in \mathbb{N}$ let $A_n = \{x : |f(x)| \leq n\}$ and define $f_n = f \cdot 1_{A_n}$. Then

$$\left| \int f_n g \right| = \left| \int f(g \cdot 1_{A_n}) \right| \leq M \|g \cdot 1_{A_n}\|_q \leq M \|g\|_q$$

for all simple functions g. Also, $\|f_n\|_p \leq M$ for all n will imply $\|f\|_p \leq M$ since $\int |f_n|^p \to \int |f|^p$ by MCT. Thus, replacing f by f_n, we may assume f is bounded.

Now let (h_k) be a sequence of simple functions which converge to f uniformly. Then

$$\left| \int f \frac{|h_k|^p}{h_k} \right| \leq M \| |h_k|^{p-1} \|_q = M \|h_k\|_p^{p/q} \to M \|f\|_p^{p/q}$$

and

$$\left| \int f \frac{|h_k|^p}{h_k} - \int |h_k|^p \right| = \int |h_k|^{p-1} |f - h_k| \to 0,$$

which together with $\int |h_k|^p \to \int |f|^p = \|f\|_p^p$ yields $\|f\|_p^p \leq M \|f\|_p^{p/q}$. Thus $\|f\|_p = \|f\|_p^{p-p/q} \leq M$. $\square$

Theorem 4.54. *Let (X, μ) be a σ-finite measure space and let $1 < p, q < \infty$ be conjugate exponents. Then $L^q(X, \mu)' \cong L^p(X, \mu)$.*

Proof.   Define $T : L^p(X, \mu) \to L^q(X, \mu)'$ by $(Tf)(g) = \int fg$. Note first that $fg \in L^1(X, \mu)$ by Hölder's inequality, so the integral is well-defined. Next,

$$\left| \int fg \right| \leq \|fg\|_1 \leq \|f\|_p \|g\|_q$$

shows that $Tf \in L^q(X, \mu)'$ and $\|Tf\| \leq \|f\|_p$. Thus T is nonexpansive, and it is clearly linear and injective. We can complete the proof by showing that every $\phi \in L^q(X, \mu)'$ satisfies $\phi = Tf$ for some $f \in L^p(X, \mu)$ with $\|f\|_p \leq \|\phi\|$.

Let $\phi \in L^q(X, \mu)'$. Suppose first that $\mu(X) < \infty$ and define $\nu(A) = \phi(1_A)$ for all measurable sets A. Then $\nu \ll \mu$, so by the Radon-Nikodym theorem there exists $f \in L^1(X, \mu)$ such that $\phi(1_A) = \nu(A) = \int f \cdot 1_A \, d\mu$ for all A. Taking linear combinations yields $\phi(g) = \int fg$ for all simple functions g. From here Lemma 4.53 implies $\|f\|_p \leq \|\phi\|$, and since Tf agrees with ϕ on all simple functions we obtain $\phi = Tf$ by the density of simple functions in $L^q(X, \mu)$.

If X is σ-finite, let w be a weight function and observe that the map $f \mapsto fw^{1/p}$ describes an isometric isomorphism between $L^p(X, \mu)$ and $L^p(X, \tilde{\mu})$, and similarly for $L^q(X, \mu)$. It therefore immediately follows from the finite case that $L^q(X, \mu)' \cong L^p(X, \mu)$ (and the pairing $\int fg\, d\mu$ is the same because $\int f w^{1/p} \cdot g w^{1/q}\, d\tilde{\mu} = \int fg \cdot w\, d\tilde{\mu} = \int fg\, d\mu$). $\qquad\square$

Since the conjugacy condition is symmetric, we can also say that $L^p(X, \mu)' \cong L^q(X, \mu)$. Thus $L^p(X, \mu) \cong L^p(X, \mu)''$. Moreover, this isomorphism is just the map $f \mapsto \hat{f}$ of Definition 3.30, so $L^p(X, \mu)$ is reflexive for $1 < p < \infty$.

4.8 Exercises

4.1. Let E be a separable Banach space and E_0 a closed subspace. Prove that E_0' is isometrically isomorphic and weak* homeomorphic to a quotient of E' by a weak* closed subspace.

4.2. Let E be a separable Banach space and E_0 a closed subspace. Prove that $(E/E_0)'$ is isometrically isomorphic and weak* homeomorphic to a weak* closed subspace of E'.

4.3. Given any norm on $\mathbf{F}^n$, show that the weak* topology on its dual is the usual topology on $\mathbf{F}^n$.

4.4. Give an example of a sequence in l^∞ which converges on each coordinate but which does not converge in the weak* topology coming from the predual l^1.

4.5. Let E be a separable normed vector space and let F be a subspace of E'. Prove that F is weak* dense in E' iff the only weak* continuous linear functional on E' which vanishes on F is the zero functional.

4.6. Let E be a separable normed vector space. We know that $[E']_1$ with the (relative) weak* topology is a compact metrizable space. Show that the map $x \mapsto \hat{x}|_{[E']_1}$ is a linear isometry from E into $C([E']_1)$.

4.7. Let E be a normed vector space and let F be a subspace of E'. Prove that any linear functional on F which is continuous for the relative weak* topology extends to a weak* continuous linear functional on E'.

4.8. Let E be a separable Banach space. Show that the map taking a closed subspace of E to the set of bounded linear functionals which vanish on it establishes a bijection between the norm closed subspaces of E and the weak* closed subspaces of E'.

4.9. Giving l^∞ the weak* topology arising from the predual l^1, identify its unital, weak* closed, self-adjoint subalgebras and its weak* closed, self-adjoint ideals.

4.10. Prove Lemma 4.15.

4.11. Show that the weak* closure of a bounded convex set in the dual of any separable normed vector space is convex.

4.12. Prove a version of Theorem 4.17 for the norm topology in a separable normed vector space.

4.13. Prove a version of Corollary 4.19 in which A and B are disjoint convex sets and A is weak* open.

4.14. Prove a version of Corollary 4.19 in which A is a weak* closed subspace and B is a single point that is not contained in A.

4.15. Prove or find a counterexample: if C is a weak* closed convex subset of l^2 that does not contain the origin then there exists $\epsilon > 0$ and $\mathbf{b} \in l^2$ such that $\Re\langle \mathbf{a}, \mathbf{b} \rangle \geq \epsilon$ for all $\mathbf{b} \in C$.

4.16. Identify $\text{ext}([l^1]_1)$.

4.17. Let E and F be separable normed vector spaces, let K be a weak* compact convex subset of E', and let $T : E' \to F'$ be a weak* continuous linear map. Prove that $T(K)$ is a weak* compact convex subset of F', and for every $g \in \text{ext}(T(K))$ there exists $f \in \text{ext}(K)$ such that $Tf = g$.

4.18. Let X be a compact metrizable space. Identify the extreme points of $[M(X)]_1$.

4.19. (Banach-Stone theorem) If X and Y are compact metrizable spaces and $T : C(X) \cong C(Y)$ is an isometric isomorphism, prove that there is a homeomorphism $\phi : Y \cong X$ and a function $g \in C(Y)$ with $|g| = 1_Y$ such that $Tf = g \cdot (f \circ \phi)$ for all $f \in C(X)$. (Use Exercise 4.18.)

4.20. Show that no clopen (simultaneously closed and open) subset of the Cantor set K can be expressed as a union of infinitely many disjoint nonempty clopen sets. Use Theorem 2.14 to prove the Riesz-Markov theorem for positive linear functionals on $C(K)$.

4.21. Let X be a countably infinite compact Hausdorff space. Prove that $C(X)' \cong l^1$.

4.22. Let (X, μ) be a measure space. Prove that $L^1(X, \mu)$ and $L^\infty(X, \mu)$ are vector spaces and $\| \cdot \|_1$ and $\| \cdot \|_\infty$ are norms.

4.23. Prove that $L^1[0, 1]$ is not isometrically isomorphic to the dual of any Banach space. (Look at extreme points.)

4.24. Let (X, μ) be a σ-finite measure space. Prove that the simple functions are uniformly dense in $L^\infty(X, \mu)$. Assuming that (X, μ) is measurably separable, show that the simple functions in $L^1(X, \mu)$ are weak* dense in $L^\infty(X, \mu)$.

4.25. Let (X, μ) be a measurably separable σ-finite measure space. Say that A *essentially contains* B if $\mu(B \setminus A) = 0$. Let S be a family of measurable subsets of X. Show that there exists a measurable subset B such that (1) each $A \in S$ is essentially contained in B, and (2) B is essentially contained in any measurable set that essentially contains every $A \in S$.

4.26. Let (X, μ) be a measurably separable σ-finite measure space. Identify the weak* closed faces of the positive part of the unit ball of $L^\infty(X, \mu)$.

4.27. Prove that the continuous functions with compact support on $\mathbf{R}^n$ are dense in $L^1(\mathbf{R}^n, m)$.

4.28. (Lusin's theorem) Let X be a compact metrizable space, let μ be a finite positive Borel measure on X, let $f \in L^\infty(X, \mu)$, and let $\epsilon > 0$. Find $g \in C(X)$ such that $\mu(\{x : f(x) \neq g(x)\}) \leq \epsilon$. (Find a sequence in $C(X)$ that converges to f in $L^1(X, \mu)$, pass to a subsequence that converges almost everywhere, and apply Exercise 2.8.)

4.29. Let (X, μ) be a measure space and let $1 \leq p < q < r \leq \infty$. Prove that $L^p(X, \mu) \cap L^r(X, \mu)$ is a Banach space with respect to the norm $\|f\| = \|f\|_p + \|f\|_r$ and that the inclusion map $L^p(X, \mu) \cap L^r(X, \mu) \hookrightarrow L^q(X, \mu)$ is bounded for this norm.

4.30. Show that $L^p(\mathbf{R})$ is separable for $1 \leq p < \infty$ and nonseparable for $p = \infty$.

4.31. Let $1 \leq p < q \leq \infty$. Show that $l^p \subseteq l^q$ and $\|\mathbf{a}\|_q \leq \|\mathbf{a}\|_p$ for all $\mathbf{a} \in l^p$.

4.32. Let $1 \leq p < q \leq \infty$. Show that $L^q([0, 1], m) \subseteq L^p([0, 1], m)$.

4.33. Let (X, μ) be a σ-finite measure space, let $1 < p, q < \infty$ be conjugate exponents, let $f : X \to \mathbf{F}$ be measurable, and suppose $fg \in L^1(X, \mu)$ for all $g \in L^q(X, \mu)$. Show that $f \in L^p(X, \mu)$. (Consider the operator $T : L^q(X, \mu) \to L^1(X, \mu)$ defined by $Tg = fg$.)

4.34. Let (X, μ) be a finite measure space and let $1 < p < \infty$. Let $f, f_n \in L^p(X, \mu)$ and suppose that $\sup \|f_n\|_p < \infty$ and $f_n \to f$ almost everywhere. Prove that $f_n \to f$ weak*. Construct an example in which $f_n \to f$ weak* but $|f - f_n| = 1_X$ for all n.

Chapter 5

Spectral Theory

5.1 Hilbert spaces

Hölder's inequality gives us a pairing between $L^p(X, \mu)$ and $L^q(X, \mu)$ where q is the conjugate exponent to p. Since 2 is conjugate to itself, this allows us to define an inner product on $L^2(X, \mu)$ by $\langle f, g \rangle = \int f \bar{g} \, d\mu$. When (X, μ) is an n-element set equipped with counting measure, the natural identification of $L^2(X, \mu)$ with $\mathbf{F}^n$ turns the integral into a sum, and we recover the usual inner product in Euclidean space. In $L^2(X, \mu)$ for general (X, μ) the norm is still related to the inner product by the formula $\|f\|^2 = \langle f, f \rangle$.

Having an inner product which gives rise to a norm in this way makes that norm extremely rigid: any vector can be taken to any other vector with the same norm via an isometric isomorphism of the space with itself, and any two infinite-dimensional, separable Banach spaces of this type are isometrically isomorphic. This implies the fact we mentioned earlier that any two closed infinite-dimensional subspaces of l^2 are isometrically isomorphic to each other, and it also implies that l^2 is isometrically isomorphic to $L^2(\mathbf{R}, m)$, to $L^2(\mathbf{R}^2, m)$, and so on.

Facts like these are hardly obvious, but they become rather easy when we approach the subject abstractly.

Definition 5.1. An *inner product* on a vector space E over $\mathbf{F}$ is a map $\langle \cdot, \cdot \rangle : E \times E \to \mathbf{F}$ satisfying the conditions

(i) $\langle v, v \rangle \geq 0$, with equality iff $v = 0$;
(ii) $\langle w, v \rangle = \overline{\langle v, w \rangle}$;
(iii) $\langle av + a'v', w \rangle = a \langle v, w \rangle + a' \langle v', w \rangle$

for all $v, v', w \in E$ and all $a, a' \in \mathbf{F}$. We define $\|v\| = \langle v, v \rangle^{1/2}$.

Combining properties (ii) and (iii) yields

$$\langle v, bw + b'w' \rangle = \bar{b}\langle v, w \rangle + \bar{b}'\langle v, w' \rangle$$

for all $v, w, w' \in E$ and $b, b' \in \mathbf{F}$.

We need to prove that the norm $\|\cdot\|$ we just defined satisfies the triangle inequality. The proof for $\mathbf{R}^n$ given in Section 3.1 works here too, but rather than repeat it, we give a formally simpler argument based on the Cauchy-Schwarz inequality.

Proposition 5.2. *(Cauchy-Schwarz inequality) Let v and w be elements of a vector space equipped with an inner product. Then $|\langle v, w \rangle| \leq \|v\|\|w\|$.*

Proof. If $w = 0$ then $\langle v, w \rangle = \langle v, 0 \cdot w \rangle = 0$ and we are done. Otherwise, evaluating the inner product of the vector $\|w\|v - \frac{\langle v,w \rangle}{\|w\|}w$ with itself yields the result $\|w\|^2\|v\|^2 - |\langle v, w \rangle|^2$. Since this quantity must be greater than or equal to zero, we infer the stated inequality. $\quad\square$

If $\|w\| = 1$ then the Cauchy-Schwarz inequality states that $|\langle v, w \rangle| \leq \|v\|$. Geometrically, this just says that the component of v in the w direction, the vector $\langle v, w \rangle w$, has length at most $\|v\|$. According to the Pythagorean theorem, the perpendicular component should have squared length $\|v\|^2 - |\langle v, w \rangle|^2$; this was the quantity, adjusted to allow w to have non-unit length, that we evaluated in the preceding proof.

Corollary 5.3. *Let v and w be elements of a vector space equipped with an inner product. Then $\|v + w\| \leq \|v\| + \|w\|$.*

Proof. We calculate:

$$\begin{aligned}
\|v + w\|^2 &= \langle v + w, v + w \rangle \\
&= \|v\|^2 + 2\Re\langle v, w \rangle + \|w\|^2 \\
&\leq \|v\|^2 + 2\|v\|\|w\| + \|w\|^2 \\
&= (\|v\| + \|w\|)^2.
\end{aligned}$$

Taking square roots yields the triangle inequality. $\quad\square$

Together with the easy observation that $\|av\| = |a|\|v\|$, this shows that any norm derived from an inner product as in Definition 5.1 really is a norm. We can now define Hilbert spaces.

Definition 5.4. A *Hilbert space* is a vector space equipped with an inner product for which the associated norm is complete.

Thus the spaces l_n^2 are finite-dimensional examples of Hilbert spaces, the space l^2 is also a Hilbert space, and indeed $L^2(X,\mu)$ for any measure space (X,μ) is a Hilbert space. We will see that, up to isometric isomorphism, there are no other examples than these. But it would be wrong to think that this means there is no value in developing an abstract theory. Using an abstract point of view can be helpful when the extra structure carried by L^2 spaces is irrelevant. Also, we may want to realize a single Hilbert space as an L^2 space in different ways.

We will call a linear bijection between Hilbert spaces that preserves the inner product a *Hilbert isomorphism*, and we will write $H_1 \cong H_2$ to indicate that there is a Hilbert isomorphism between H_1 and H_2. This slightly conflicts with our earlier use of the symbol $\cong$ in the Banach space setting to indicate isometric isomorphism, but the conflict is not substantive because any Hilbert isomorphism is automatically an isometric isomorphism — this follows from the fact that the norm is derived from the inner product — and conversely, any isometric isomorphism between two Hilbert spaces must be a Hilbert isomorphism (see Exercise 5.17).

The key geometric concept that we have in Hilbert spaces but not in general Banach spaces is orthogonality.

Definition 5.5. Let H be a Hilbert space. We say that $v, w \in H$ are *orthogonal*, and write $v \perp w$, if $\langle v, w \rangle = 0$. For closed subspaces E and F of H we write $E \perp F$, and say that E and F are orthogonal, when $v \perp w$ for all $v \in E$ and $w \in F$. The *orthocomplement* of E is the closed subspace $E^\perp = \{v \in H : v \perp w \text{ for all } w \in E\}$.

$E^\perp$ is a subspace by the linearity of inner products expressed in Definition 5.1 (iii). It is closed because $v_n \to 0$ implies $\langle v_n, w \rangle \to 0$ by Cauchy-Schwarz (and hence $v_n \to v$ implies $\langle v_n, w \rangle \to \langle v, w \rangle$).

We note a more general consequence of the Cauchy-Schwarz inequality: if $v_n \to v$ and $w_n \to w$ then $\langle v_n, w_n \rangle \to \langle v, w \rangle$. This can be seen by the computation

$$|\langle v, w \rangle - \langle v_n, w_n \rangle| = |\langle v - v_n, w \rangle + \langle v_n, w - w_n \rangle|$$
$$\leq \|v - v_n\|\|w\| + \|v_n\|\|w - w_n\|.$$

The next result shows that when subspaces are orthogonal to each other they behave like the terms of an l^2 direct sum (Definition 3.21 (b)). By Proposition 3.22, any l^2 direct sum of Hilbert spaces will be complete, and on any such direct sum $\bigoplus^2 H_n$ we have a natural inner product defined by $\langle \mathbf{v}, \mathbf{w} \rangle = \sum \langle v_n, w_n \rangle$, taking the inner product of the nth components in the nth Hilbert space. This series is summable by the estimate $|\langle v_n, w_n \rangle| \leq \|v_n\| \|w_n\| \leq \frac{1}{2}(\|v_n\|^2 + \|w_n\|^2)$, and one easily checks that it satisfies the axioms for an inner product. Thus the l^2 direct sum of a family of Hilbert spaces is itself a Hilbert space. Since the l^2 direct sum is the only direct sum we will use in this chapter, we will drop the superscript and just write it as $\bigoplus$.

Proposition 5.6. *Let (E_n) be a finite or infinite sequence of closed subspaces of a Hilbert space H which satisfy $E_m \perp E_n$ for $m \neq n$, and let E be their closed span. Then the map $(v_n) \mapsto \sum v_n$ defines a Hilbert isomorphism between $\bigoplus E_n$ and E.*

Proof. First define $U\mathbf{v} = \sum v_n$ for any sequence $\mathbf{v} = (v_n) \in \bigoplus E_n$ which has only finitely many nonzero terms. There are no convergence issues on these sequences, and U preserves the inner product because $v_m \perp w_n$ for $m \neq n$ entails that

$$\left\langle \sum v_n, \sum w_n \right\rangle = \sum \langle v_n, w_n \rangle = \langle \mathbf{v}, \mathbf{w} \rangle.$$

It follows that U is isometric, and it therefore extends by continuity to a Hilbert isomorphism from $\bigoplus E_n$ into E. Moreover, the extended map satisfies the same formula $U\mathbf{v} = \sum v_n$ for all $\mathbf{v} \in \bigoplus E_n$ by continuity. It maps onto E because its range contains each E_n and it is isometric, hence its range is closed. $\square$

For example, let (X, μ) be a measure space, let (A_n) be a sequence of disjoint measurable subsets of X, and let $A = \bigcup A_n$. Then for each n the set of L^2 functions supported on A_n constitutes a closed subspace of $L^2(X, \mu)$ which can be identified with $L^2(A_n, \mu|_{A_n})$, and for distinct values of n these subspaces are orthogonal. The set of L^2 functions supported on A is isomorphic to their l^2 direct sum by the map $f \mapsto (f|_{A_n})$. Thus $L^2(A, \mu|_A) \cong \bigoplus L^2(A_n, \mu|_{A_n})$. In the sequel we will identify $L^2(A, \mu|_A)$ with a subspace of $L^2(X, \mu)$ without comment.

Another consequence of Proposition 5.6 is that the closed span of a sequence of mutually orthogonal one-dimensional subspaces of a Hilbert space H is Hilbert isomorphic to l^2. Under this isomorphism the vectors $\mathbf{e}^n \in l^2$ correspond to vectors $e_n \in H$ with the characteristic properties that $\|e_n\| = 1$ for all n and $e_m \perp e_n$ for all $m \neq n$. We say that such vectors are *orthonormal*.

Lemma 5.7. *Let $e_1, \ldots, e_n$ be orthonormal vectors in a Hilbert space H and let $v \in H$. Then we can write $v = v_1 + v_2$ with*

$$v_1 = \langle v, e_1 \rangle e_1 + \cdots + \langle v, e_n \rangle e_n \in \mathrm{span}\{e_1, \ldots, e_n\}$$

and $v_2 \in \text{span}\{e_1, \ldots, e_n\}^{\perp}$.

Proof. It is obvious that v_1 as defined belongs to the span of $e_1, \ldots, e_n$. Define $v_2 = v - v_1$; then $v = v_1 + v_2$ is immediate and we just need to verify that $v_2 \in \text{span}\{e_1, \ldots, e_n\}^{\perp}$. It is enough to check that $v_2 \perp e_i$, i.e., that $\langle v, e_i \rangle = \langle v_1, e_i \rangle$, for $1 \leq i \leq n$. But since $e_i \perp e_j$ for $i \neq j$, taking the inner product of v_1 with e_i yields only the term

$$\langle \langle v, e_i \rangle e_i, e_i \rangle = \langle v, e_i \rangle \cdot \|e_i\|^2 = \langle v, e_i \rangle.$$

So $\langle v - v_1, e_i \rangle = 0$, as desired. $\square$

An *orthonormal basis* of a Hilbert space H is an orthonormal set of vectors whose span is dense in H.

Theorem 5.8. *Let H be a separable Hilbert space. Then H has a countable orthonormal basis.*

Proof. Let (v_n) be a finite or infinite sequence of linearly independent vectors whose closed span equals H. Such a sequence can be found by starting with a dense sequence and removing each vector which is a linear combination of previous vectors in the sequence.

Let $e_1 = v_1$. Given $e_1, \ldots, e_{n-1}$, use the lemma to write $v_n = w_n + e_n$ where $w_n \in \text{span}\{e_1, \ldots, e_{n-1}\}$ and $e_n \in \text{span}\{e_1, \ldots, e_{n-1}\}^{\perp}$. Then e_n must be nonzero because v_n is not a linear combination of $v_1, \ldots, v_{n-1}$, and we inductively have $\text{span}\{e_1, \ldots, e_n\} = \text{span}\{v_1, \ldots, v_n\}$.

Thus the e_n are mutually orthogonal nonzero vectors whose span is dense in H. It follows that the vectors $\frac{1}{\|e_n\|} e_n$ form an orthonormal basis of H. $\square$

Our use of the term "basis" varies from its use in linear algebra, where a set is said to be a basis if it is linearly independent and it spans the whole space. We require only that the span be dense. In finite dimensions this makes no difference, so we can define the dimension of H, $\dim(H)$, to equal $n \in \mathbf{N}$ if there is an orthonormal basis whose cardinality is n, and we know from linear algebra that this quantity is well-defined, i.e., any other basis must have the same number of elements. If there is no finite orthonormal basis then we set $\dim(H) = \infty$.

The existence of orthonormal bases has a stunning consequence. We introduce here the alternative notation l_{∞}^2 for l^2; this allows us to uniformly write l_n^2 for $0 \leq n \leq \infty$.

Corollary 5.9. *Let H be a separable Hilbert space. Then $H \cong l_n^2$ where $n = \dim H$.*

Proof. Let (e_n) be an orthonormal basis of H. Then the one-dimensional subspaces $\mathbf{F} \cdot e_n$ of H spanned by the basis vectors form a countable family of mutually orthogonal closed subspaces whose span is dense in H. So by Proposition 5.6 the map taking $\sum a_n e_n$ to (a_n) defines a Hilbert isomorphism between H and the l^2 direct sum of $\dim H$ copies of $\mathbf{F}$, which can be identified with l_n^2. $\qquad\square$

This corollary makes the structure of abstract Hilbert spaces quite transparent, because in effect it means that we can treat any orthonormal basis as if it were the standard orthonormal basis of l_n^2. For example, the following computations are immediate consequences.

Corollary 5.10. *Let (e_n) be an orthonormal basis of a separable Hilbert space and let $v, w \in H$. Then*

$$\|v\|^2 = \sum |\langle v, e_n \rangle|^2$$

and

$$\langle v, w \rangle = \sum \langle v, e_n \rangle \langle e_n, w \rangle.$$

The point is that in l_n^2 the components of an element $\mathbf{a}$ are given by the inner products $\langle \mathbf{a}, \mathbf{e}^n \rangle$. Thus we have $\|\mathbf{a}\|^2 = \sum |\langle \mathbf{a}, \mathbf{e}^n \rangle|^2$ and $\langle \mathbf{a}, \mathbf{b} \rangle = \sum \langle \mathbf{a}, \mathbf{e}^n \rangle \langle \mathbf{e}^n, \mathbf{b} \rangle$, and these formulas transfer to arbitrary orthonormal bases by the previous corollary.

Example 5.11. $L^2([0,1], m)$ is a separable, infinite-dimensional Hilbert space, so it must be Hilbert isomorphic to l^2.

If $\mathbf{F} = \mathbf{C}$ then $L^2([0,1], m)$ has a natural orthonormal basis consisting of the functions $\exp(2\pi i n t)$ for $n \in \mathbf{Z}$, which gives rise to a natural Hilbert isomorphism with $l^2(\mathbf{Z})$. These functions are orthogonal by an elementary computation, and we know that their span is uniformly dense in $C([0,1])$ (Example 3.53). Since uniform convergence in $C([0,1])$ implies convergence in $L^2([0,1], m)$, this entails that the closed span of the functions $\exp(2\pi i n t)$ in $L^2([0,1], m)$ contains $C([0,1])$, and it is therefore all of $L^2([0,1], m)$ by Proposition 4.51 (b).

This Hilbert isomorphism between $L^2([0,1], m)$ and $l^2(\mathbf{Z})$ takes a function $f \in L^2([0,1], m)$ to $\sum a_n \mathbf{e}^n$ where $a_n = \langle f, \exp(2\pi i n t) \rangle$, i.e., (a_n) is the Fourier series of f. The two formulas given by Corollary 5.10 are known as Parseval's theorem in this context.

Similarly, $L^2(\mathbf{R}, m)$ must be Hilbert isomorphic to l^2, but here there is no really obvious standard orthonormal basis to use. One natural choice is to orthonormalize the functions $e^{-t^2}t^n$ for $n = 0, 1, 2, \ldots$ using the procedure described in Theorem 5.8; this results in a sequence of functions of the form $e^{-t^2}H_n(t)$ where H_n is a polynomial of degree n. These are known as the *Hermite polynomials*.

The key technical fact about closed subspaces of a Hilbert space is that the sum of any closed subspace and its orthocomplement equals the whole space. Using the preceding material we can prove this result fairly easily.

Lemma 5.12. *Let E be a closed subspace of a separable Hilbert space H. Then $H = E + E^{\perp}$.*

Proof. Let $v \in H$. We must express it as the sum of a vector in E and a vector in $E^{\perp}$. Recall the Pythagorean theorem: if $v \perp w$ then $\|v + w\|^2 = \|v\|^2 + \|w\|^2$. In particular, $\|v + w\| \geq \|v\|$.

Now E is itself a separable Hilbert space, so by Theorem 5.8 it has an orthonormal basis (e_n). If E is finite-dimensional then we are done by Lemma 5.7. If it is infinite-dimensional, for each n let $a_n = \langle v, e_n \rangle$, and define $v_n = a_1 e_1 + \cdots + a_n e_n$ as in Lemma 5.7. Then

$$|a_1|^2 + \cdots + |a_n|^2 = \|v_n\|^2 \leq \|v\|^2$$

by Corollary 5.10 and the Pythagorean theorem, so the series $\sum a_n e_n$ converges in $E \cong l^2$.

Let $w = \sum a_n e_n$. Then $w \in E$, and $v - w = v - \lim v_n \in E^{\perp}$ since $e_n \perp v - v_m$ for $m \geq n$. Thus we have expressed v as the sum of a vector in E and a vector in $E^{\perp}$. $\square$

This seemingly modest result has a powerful consequence.

Theorem 5.13. *(Riesz representation theorem) Let H be a Hilbert space. Then every $w \in H$ gives rise to a bounded linear functional $v \mapsto \langle v, w \rangle$ with norm equal to $\|w\|$, and every bounded linear functional on H has this form.*

Proof. It follows immediately from the definition of an inner product that the map $v \mapsto \langle v, w \rangle$ is a linear functional on H. It is bounded, with norm at most $\|w\|$, by the Cauchy-Schwarz inequality, and taking $v = w$ shows that its norm is exactly $\|w\|$.

Now let $f : H \to \mathbf{F}$ be any bounded linear functional on H. If $f = 0$ then it is given by taking inner products against $0 \in H$. Otherwise its kernel is a closed codimension 1 subspace of H, so by Lemma 5.12 the orthocomplement of $\ker f$ must be a one-dimensional subspace, say $\mathbf{F} \cdot w_0$. We can assume $\|w_0\| = 1$.

Let $w = \overline{f(w_0)}w_0$; we will show that $f(v) = \langle v, w \rangle$ for all $v \in H$. First, any $v \in \ker f$ is orthogonal to w_0, and hence $f(v) = 0 = \langle v, w \rangle$ for all such v. We also have $\langle w_0, w \rangle = \langle w_0, \overline{f(w_0)}w_0 \rangle = f(w_0)$, and since every element of H can be written in the form $v + aw_0$ for some $v \in \ker f$ and $a \in \mathbf{F}$, it follows that f is given by taking inner products against w. $\square$

This result essentially recovers Theorem 4.54 in the case $p = 2$.

5.2   Hilbert bundles

A Hilbert bundle is an object which combines the structure of a measure space with the structure of a Hilbert space.

Definition 5.14. Let (X, μ) be a measure space. A *measurable Hilbert bundle* over X is a set of the form

$$\mathcal{H} = \bigcup (X_n \times H_n)$$

where the X_n are countably many disjoint measurable subsets whose union is X and the H_n are separable Hilbert spaces of dimension at least 1.

We can think of X as the base space and the Hilbert spaces H_n as the fibers over the points of X. Since we are working in a measurable, not topological, setting, our bundles are quite simple: the dimension of the fiber can vary, so we allow countably many different fibers, but on the region corresponding to a single fiber dimension the bundle is just a cartesian product. The prohibition on zero-dimensional fibers is convenient because it prevents a kind of degeneracy that we would otherwise occasionally have to explicitly exclude.

We will associate a Hilbert space to each measurable Hilbert bundle, intuitively the space of L^2 sections of the bundle. This will provide us with a generalization of $L^2(X, \mu)$ spaces which is well suited to doing spectral theory.

The basic building block of this construction is given in the following definition.

Definition 5.15. Let (X, μ) be a measure space and let H be a separable Hilbert space. A function $\eta : X \to H$ is *weakly measurable* if for each $w \in H$ the function $x \mapsto \langle \eta(x), w \rangle$ is measurable from X to $\mathbf{F}$.

We define $L^2(X, \mu; H)$ to be the set of all weakly measurable functions $\eta : X \to H$ for which $\|\eta\| = (\int \|\eta(x)\|^2 \, d\mu)^{1/2}$ is finite, identifying functions with agree almost everywhere, and we define an inner product on

$L^2(X, \mu; H)$ by

$$\langle \eta, \xi \rangle = \int \langle \eta(x), \xi(x) \rangle \, d\mu.$$

Proposition 5.16. *Let (X, μ) be a measure space and let H be a separable Hilbert space. Then $L^2(X, \mu; H)$ is a Hilbert space.*

Proof. Several things need to be checked. First, note that the function $x \mapsto \|\eta(x)\|^2$ is measurable because Corollary 5.10 allows us to express it as a sum of positive measurable functions, $\|\eta(x)\|^2 = \sum |\langle \eta(x), e_n \rangle|^2$, where (e_n) is an orthonormal basis of H. If η and ξ both belong to $L^2(X, \mu; H)$ then the second part of Corollary 5.10 shows that $\langle \eta(x), \xi(x) \rangle$ is measurable, and the pointwise inequality

$$|\langle \eta(x), \xi(x) \rangle| \le \|\eta(x)\| \|\xi(x)\| \le \frac{1}{2}(\|\eta(x)\|^2 + \|\xi(x)\|^2)$$

shows that it is integrable. Thus the inner product on $L^2(X, \mu; H)$ is well-defined. Verifying that it satisfies the conditions of Definition 5.1 is straightforward.

The proof of completeness mimics the proof of completeness of $L^2(X, \mu)$ (Proposition 4.50). Suppose (η_n) is an absolutely summable series in $L^2(X, \mu; H)$. Then

$$\left(\int \left(\sum_{n=1}^{N} \|\eta_n(x)\| \right)^2 d\mu \right)^{1/2} \le \sum_{n=1}^{N} \left(\int \|\eta_n(x)\|^2 \, d\mu \right)^{1/2} \le \sum_{n=1}^{\infty} \|\eta_n\| < \infty,$$

using Minkowski's inequality in $L^2(X, \mu)$. Taking the limit $N \to \infty$ and applying MCT shows that the function $g : x \mapsto (\sum_{n=1}^{\infty} \|\eta_n(x)\|)^2$ is integrable, so it must be finite almost everywhere and we can therefore define the sum $\eta(x) = \sum_{n=1}^{\infty} \eta_n(x)$ in H for almost every x. The resulting function is weakly measurable because for any $w \in H$ the function $x \mapsto \langle \eta(x), w \rangle$ is almost everywhere the pointwise limit of the measurable functions $x \mapsto \langle \sum_{n=1}^{N} \eta_n(x), w \rangle$. It belongs to $L^2(X, \mu; H)$ because $\|\eta(x)\|^2 \le g(x)$ for almost every x. Finally, $\|\eta - \sum_{n=1}^{N} \eta_n\| = \|\sum_{n=N+1}^{\infty} \eta_n\|$ converges to zero by DCT because $\|\sum_{n=N+1}^{\infty} \eta_n(x)\|^2 \le g(x)$ almost everywhere and converges pointwise almost everywhere to zero. $\qquad \square$

It may be instructive to note that $L^2(\mathbf{N}, \mu_{\text{count}}; H)$ is isometrically isomorphic to the direct sum of a sequence of copies of H. A typical element of $L^2(\mathbf{N}, \mu_{\text{count}}; H)$ is a function from $\mathbf{N}$ into H, and this corresponds to a sequence of elements of H in an obvious way. The membership condition

$\sum \|v_n\|^2 < \infty$ is the same in the two cases, and the definitions of inner product and norm also agree. Thus, we can think of the general construction of $L^2(X, \mu; H)$ as a sort of "measurable direct sum".

Direct sums can also illuminate the structure of $L^2(X, \mu; H)$ in a different way.

Proposition 5.17. *Let (X, μ) be a measure space and let H be a separable Hilbert space with orthonormal basis (e_n). Then the map $(f_n) \mapsto \sum f_n \cdot e_n$ defines a Hilbert isomorphism between $\bigoplus L^2(X, \mu)$ and $L^2(X, \mu; H)$, taking the direct sum of $\dim H$ copies of $L^2(X, \mu)$.*

Proof.   For each n define $E_n = \{f \cdot e_n : f \in L^2(X, \mu)\} \subseteq L^2(X, \mu; H)$. Then E_n is Hilbert isomorphic to $L^2(X, \mu)$ and $E_m \perp E_n$ for $m \neq n$. The result now follows from Proposition 5.6 provided we can show that the span of the E_n is dense in $L^2(X, \mu; H)$.

By Lemma 5.12 it will suffice to show that the orthocomplement of the span of the E_n is zero. So suppose $\eta \in L^2(X, \mu; H)$ is orthogonal to every element of the form $f \cdot e_n$. Taking $f_n(x) = \langle \eta(x), e_n \rangle$ then yields

$$\int |\langle \eta(x), e_n \rangle|^2 = \int \langle \eta(x), f_n(x) \cdot e_n \rangle = \langle \eta, f_n \cdot e_n \rangle = 0.$$

So $\langle \eta(x), e_n \rangle = 0$ for almost every x, for all n. This shows that $\eta(x) = 0$ almost everywhere, as desired.   □

We now define the Hilbert space associated to a measurable Hilbert bundle.

Definition 5.18. Let (X, μ) be a measure space and let $\mathcal{H} = \bigcup (X_n \times H_n)$ be a measurable Hilbert bundle over X. The *space of L^2 sections* of $\mathcal{H}$ is the Hilbert space

$$L^2(X, \mu; \mathcal{H}) = \bigoplus L^2(X_n, \mu|_{X_n}; H_n).$$

Thus, a typical element of $L^2(X, \mu; \mathcal{H})$ is a family of functions $\eta_n : X_n \to H_n$. We can think of it as a single function on X whose restriction to X_n takes values in H_n. Under this interpretation we can use the same formulas $\|\eta\| = (\int \|\eta(x)\|^2 \, d\mu)^{1/2}$ and $\langle \eta, \xi \rangle = \int \langle \eta(x), \xi(x) \rangle \, d\mu$ that we gave for $L^2(X, \mu; H)$.

Since all Hilbert spaces of the same dimension are isomorphic, there is no loss in generality in assuming that $H_n = l_n^2$, provided we include the value $n = \infty$. (Recall that we use l_∞^2 as an alternative notation for

l^2.) Thus we can express X as a union of disjoint measurable subsets X_n, $1 \leq n \leq \infty$, and take $\mathcal{H}$ to be the bundle

$$\mathcal{H} = \bigcup_{1 \leq n \leq \infty} (X_n \times l_n^2).$$

Then, regarding l_n^2 as sitting inside l_∞^2 as the span of the first n standard basis vectors, we can identify $L^2(X, \mu; \mathcal{H})$ with the set of functions $\eta \in L^2(X, \mu; l^2)$ which, for each n, satisfy $\langle \eta(x), \mathbf{e}^m \rangle = 0$ for almost every $x \in X_n$ and every $m > n$.

The underlying measure space (X, μ) imparts a rich structure on the Hilbert space $L^2(X, \mu; \mathcal{H})$. Namely, for every measurable subset $A \subseteq X$ we have a closed subspace consisting of those L^2 sections supported on A. As with $L^2(X, \mu)$, in the sequel we will identify the L^2 sections supported on A with $L^2(A, \mu|_A; \mathcal{H}|_A)$ without comment. This association of measurable subsets of X to closed subspaces of $L^2(X, \mu; \mathcal{H})$ has nice properties which are characterized abstractly in the following definition.

Definition 5.19. Let X be a measurable space and let H be a Hilbert space. An *(H-valued) spectral measure* on X is a function λ taking measurable subsets of X to closed subspaces of H which satisfies the conditions

(i) $\lambda(\emptyset) = \{0\}$, $\lambda(X) = H$;
(ii) $\lambda(A) \perp \lambda(B)$;
(iii) $\lambda(\bigcup A_n) = \overline{\mathrm{span}}(\bigcup \lambda(A_n))$

for any disjoint measurable subsets $A, B \subseteq X$ and any sequence of disjoint measurable subsets $A_n \subseteq X$.

If $\mathcal{H}$ is a measurable Hilbert bundle over X, then the function $\lambda : A \mapsto L^2(A, \mu|_A; \mathcal{H}|_A)$ taking A to the set of functions in $L^2(X, \mu; \mathcal{H})$ supported on A satisfies the properties of a spectral measure on X. We call this λ the *standard $L^2(X, \mu; \mathcal{H})$-valued spectral measure* on X.

Our goal in the rest of this section is to prove that every abstract spectral measure can be realized in the preceding form. The appropriate notion of isomorphism here is the following. An *equivalence* between H- and H'-valued spectral measures λ and λ' on a measurable space X is a Hilbert isomorphism $U : H \cong H'$ such that $\lambda'(A) = U(\lambda(A))$ for all measurable $A \subseteq X$, and λ and λ' are *equivalent* if there exists an equivalence between them.

Before continuing, we mention an easy observation that can be helpful in establishing equivalence.

Lemma 5.20. *Let λ and λ' be H- and H'-valued spectral measures on a measurable space X, and let $U : H \cong H'$ be a Hilbert isomorphism. Suppose $U(\lambda(A)) \subseteq \lambda'(A)$ for all measurable $A \subseteq X$. Then U is an equivalence between λ and λ'.*

Proof. Fix $A \subseteq X$; we want to show that $U(\lambda(A)) = \lambda'(A)$. Let $w \in \lambda'(A)$. Since U is an isomorphism there exists $v \in H$ such that $Uv = w$, and by Lemma 5.12 we can write $v = v_1 + v_2$ where $v_1 \in \lambda(A)$ and $v_2 \in \lambda(A)^{\perp} = \lambda(A^c)$. Then we have $w = Uv_1 + Uv_2$. But $w, Uv_1 \in \lambda'(A)$ and $Uv_2 \in \lambda'(A^c) = \lambda'(A)^{\perp}$, so we must have $v_2 = 0$ and $w = Uv_1$. We conclude that $\lambda'(A) = U(\lambda(A))$. $\qquad\square$

For any closed subspace E of a Hilbert space H, define a map $P_E : H \to H$ by letting $P_E v$ be the component of v in E, i.e., $P_E v = v_1$ where $v = v_1 + v_2$ with $v_1 \in E$ and $v_2 \in E^{\perp}$. We call P_E the *projection* onto E. Note that the decomposition $v = v_1 + v_2$ is unique because if $v = v'_1 + v'_2$ were another such decomposition then $v_1 + v_2 = v'_1 + v'_2$ would imply $v_1 - v'_1 = v'_2 - v_2$, which entails that $v_1 = v'_1$ and $v_2 = v'_2$ since $E \cap E^{\perp} = \{0\}$.

Below we will use the property that

$$\langle P_{\lambda(A)}v, P_{\lambda(B)}w \rangle = \langle P_{\lambda(A \cap B)}v, P_{\lambda(A \cap B)}w \rangle$$

for all measurable $A, B \subseteq X$. This holds because $P_{\lambda(A)}v$ can be further decomposed into a component in $\lambda(A \cap B)$ and a component in $\lambda(A \setminus B)$, while $P_{\lambda(B)}w$ can be decomposed into a component in $\lambda(A \cap B)$ and a component in $\lambda(B \setminus A)$, and after expanding out $\langle P_{\lambda(A)}v, P_{\lambda(B)}w \rangle$ accordingly the only nonzero term is the inner product between $P_{\lambda(A \cap B)}v$ and $P_{\lambda(A \cap B)}w$.

Now say that $v \in H$ is *cyclic* for an H-valued spectral measure λ if the span of the vectors $P_{\lambda(A)}v$, for $A \subseteq X$ measurable, is dense in H. There need not exist any cyclic vectors, but if there do then the spectral measure has a particularly simple form.

Lemma 5.21. *Let X be a measurable space, H a Hilbert space, and λ an H-valued spectral measure on X. Suppose there is a cyclic vector v for λ. Then $\mu(A) = \|P_{\lambda(A)}v\|^2$ is a finite measure on X and λ is equivalent to the standard $L^2(X, \mu)$-valued spectral measure on X.*

Proof. The condition $\mu(\emptyset) = 0$ is trivial. Let (A_n) be a sequence of disjoint measurable subsets of X. Then the subspaces $\lambda(A_n)$ are mutually orthogonal, and their span is dense in $\lambda(\bigcup A_n)$. It follows from Proposition 5.6 that $\|P_{\lambda(\bigcup A_n)}v\|^2 = \sum \|P_{\lambda(A_n)}v\|^2$. This shows that μ is countably additive. So μ is a measure. It is finite because $\mu(X) = \|P_H v\|^2 = \|v\|^2$.

Define a map U from the simple functions in $L^2(X, \mu)$ into H by setting $Uh = \sum a_i P_{\lambda(A_i)} v$ for a simple function h whose standard form is $h = \sum a_i 1_{A_i}$. Then U preserves inner products by the computation

$$\left\langle \sum_i a_i P_{\lambda(A_i)} v, \sum_j b_j P_{\lambda(B_j)} v \right\rangle = \sum_{i,j} a_i \bar{b}_j \| P_{\lambda(A_i \cap B_j)} v \|^2$$

$$= \int \sum_i a_i 1_{A_i} \cdot \overline{\sum_j b_j 1_{B_j}} \, d\mu.$$

Note that this computation does not depend on the simple functions being in standard form. So if $\sum a_i 1_{A_i} = \sum b_j 1_{B_j}$ are two expressions of the same simple function, the preceding shows that the inner product of $\sum a_i P_{\lambda(A_i)} v - \sum b_j P_{\lambda(B_j)} v$ with itself is zero, and therefore that $\sum a_i P_{\lambda(A_i)} v = \sum b_j P_{\lambda(B_j)} v$. We can infer from this that the formula $Uh = \sum a_i P_{\lambda(A_i)} v$ is well-defined without requiring h to be in standard form. It now follows that U is linear, and that U extends by continuity to a Hilbert isomorphism between $L^2(X, \mu)$ and H.

Any simple function h supported on a set $A \subseteq X$ satisfies $Uh \in \lambda(A)$, so by continuity U takes $L^2(A, \mu|_A)$ into $\lambda(A)$. Then according to Lemma 5.20, U is an equivalence between the standard $L^2(X, \mu)$-valued spectral measure on X and λ. $\quad\square$

Since not every spectral measure has a cyclic vector, the situation is more complicated than this in general. However, we will show that we can always decompose a given spectral measure λ into a family of spectral measures λ_n with cyclic vectors v_n. Then we can apply Lemma 5.21 to each λ_n, and intuitively we should be able to add up the resulting $L^2(X, \mu_n)$ spaces to get the L^2 sections of a measurable Hilbert bundle. Because the measures μ_n might be supported on different sets, some regions of X may get counted multiple or even infinitely many times. This should correspond to the structure of the Hilbert bundle we seek. We now have to make this idea rigorous. The main step is the following lemma.

Any countable family of H_n-valued spectral measures λ_n on a single measurable space X generates a natural $\bigoplus H_n$-valued spectral measure λ defined by $\lambda(A) = \bigoplus \lambda_n(A) \subseteq \bigoplus H_n$. We will refer to this λ as the direct sum of the λ_n.

Lemma 5.22. *Let (μ_n) be a finite or infinite sequence of finite measures on a measurable space X. Then the direct sum of the standard $L^2(X, \mu_n)$-valued spectral measures on X is equivalent to the standard $L^2(X, \mu; \mathcal{H})$-*

valued spectral measure for some finite measure μ and some measurable Hilbert bundle $\mathcal{H}$ over X.

Proof. Let $\mu = \sum \frac{1}{2^n \mu_n(X)} \mu_n$ and observe that $\mu_n \ll \mu$ for all n.

We construct the desired equivalence in a series of stages. By the Radon-Nikodym theorem, for each n we can find $f_n \in L^1(X, \mu)$ such that $d\mu_n = f_n\, d\mu$. Let A_n be the support of f_n. Then the map $f \mapsto \sqrt{f_n} \cdot f$ defines a Hilbert isomorphism between $L^2(X, \mu_n)$ and $L^2(A_n, \mu|_{A_n})$ which implements an equivalence between their respective associated spectral measures. Taking direct sums yields an equivalence between the spectral measures associated to $\bigoplus L^2(X, \mu_n)$ and $\bigoplus L^2(A_n, \mu|_{A_n})$.

Next, we can rearrange the sets A_n to ensure that they are decreasing. Given two sets A and B, call the operation of replacing A by $A \cup B$ and B by $B \cap A$ a *transfer of excess* from B to A. Rearrange the sets A_n as follows. In the first round, successively transfer excess from A_k to A_1 with $k = 2, 3, \ldots$. In the second round, successively transfer excess from A_k to A_2 with $k = 3, 4, \ldots$. Continue in this fashion. The final result is a sequence (B_n) which is now decreasing and whose associated Hilbert space $\bigoplus L^2(B_n, \mu|_{B_n})$ is Hilbert isomorphic to the original space $\bigoplus L^2(A_n, \mu|_{A_n})$ in a way that preserves the associated spectral measures. This can be seen by creating an array of sets A_n^k with, for $k > n$, $A_n^k =$ the portion of A_k transferred to A_n, and $A_k^k = A_k$ minus all of the portions that get transferred to preceding A_k's. Then $A_k = \bigcup_{n \leq k} A_n^k$ (all the portions that were removed plus the portion that was not) and $B_n = \bigcup_{k \geq n} A_n^k$ (what is left of A_n after transfers to preceding sets, plus all the portions that get transferred in). This shows how to break up and rearrange one Hilbert space to get the other.

Now for each $n \in \mathbf{N}$ let $X_n = B_n \setminus B_{n+1}$, and let $X_\infty = \bigcap B_n$. Let $\mathcal{H}$ be the corresponding measurable Hilbert bundle with $H_n = l_n^2$ for $1 \leq n \leq \infty$. Recall that we can identify $L^2(X, \mu; \mathcal{H})$ with the set of $\eta \in L^2(X, \mu; l^2)$ whose restriction to X_n takes values in l_n^2. So the map $U : \bigoplus L^2(B_n, \mu|_{B_n}) \to L^2(X, \mu; \mathcal{H})$ which takes (f_n) to $\sum f_n \cdot \mathbf{e}^n$ is the desired final Hilbert isomorphism that implements an equivalence between the associated spectral measures.   □

The next lemma will be used to decompose a spectral measure into cyclic components.

Lemma 5.23. *Let H be a Hilbert space, let E be a closed subspace of H, and let λ be an H-valued spectral measure on a measurable space X.*

Suppose that

$$E = (E \cap \lambda(A)) + (E \cap \lambda(A^c))$$

for all measurable $A \subseteq X$. *Then* $\lambda_E(A) = E \cap \lambda(A)$ *defines an* E-*valued spectral measure on* X.

Proof. First we establish the following claim about closed subspaces: if F is any closed subspace of H such that $E = (E \cap F) + (E \cap F^\perp)$ then we also have $F = (E \cap F) + (E^\perp \cap F)$. It is clear that F always contains the right side. For the reverse inclusion, let $v \in F$. Treating F as a Hilbert space and $E \cap F$ as a closed subspace, we can write $v = v_1 + v_2$ with $v_1 \in E \cap F$ and $v_2 \in F \cap (E \cap F)^\perp$. Then v_2 is orthogonal to every vector in $E \cap F$, and it is also orthogonal to every vector in $E \cap F^\perp$ since it belongs to F. Thus by hypothesis it is orthogonal to every vector in E, i.e., we have $v_2 \in E^\perp \cap F$. This proves the claim.

It is easy to see that λ_E satisfies conditions (i) and (ii) of Definition 5.19. For condition (iii), let (A_n) be a sequence of disjoint measurable subsets of X and let $v \in \lambda_E(\bigcup A_n)$; we must show that $v \in \overline{\text{span}}(\bigcup \lambda_E(A_n))$. Write $v = \sum v_n$ with $v_n \in \lambda(A_n)$ for all n, and use the claim to decompose each v_n as $v_n = v_n^1 + v_n^2$ with $v_n^1 \in E \cap \lambda(A_n)$ and $v_n^2 \in E^\perp \cap \lambda(A_n)$. Then $\sum v_n^1 + \sum v_n^2 = v$, with the first sum taking place in E and the second sum taking place in $E^\perp$. Since $v \in E$, it follows that $\sum v_n^1 = v$ (and in fact that every term of the second sum is zero). This shows that $\lambda_E(A) = \overline{\text{span}}(\bigcup \lambda_E(A_n))$. □

Theorem 5.24. *Let* H *be a separable Hilbert space. Then any* H-*valued spectral measure on a measurable space* X *is equivalent to the standard* $L^2(X, \mu; \mathcal{H})$-*valued spectral measure on* X, *for some finite measure* μ *and some measurable Hilbert bundle* $\mathcal{H}$ *over* X.

Proof. Let X be a measurable space and let λ be an H-valued spectral measure on X. Let (e_n) be an orthonormal basis of H and construct corresponding orthogonal closed subspaces E_n of H as follows. Start by setting $e_1' = e_1$ and letting E_1 be the closed span of the vectors $P_{\lambda(A)}e_1'$, with A ranging over the measurable subsets of X. Having constructed $E_1, \ldots, E_{n-1}$, write $e_n = v_n + e_n'$ where $v_n \in E_1 + \cdots + E_{n-1}$ and $e_n' \in (E_1 + \cdots + E_{n-1})^\perp$. Then let E_n be the closed span of the vectors $P_{\lambda(A)}e_n'$ for $A \subseteq X$.

Discard any E_n which equal $\{0\}$. We claim that $E_m \perp E_n$ for $m \neq n$. Suppose $m < n$. Then it will suffice to show that $P_{\lambda(A)}e_m' \perp P_{\lambda(B)}e_n'$ for any measurable $A, B \subseteq X$. But we have

$$\langle P_{\lambda(A)}e'_m, P_{\lambda(B)}e'_n \rangle = \langle P_{\lambda(A \cap B)}e'_m, P_{\lambda(A \cap B)}e'_n \rangle = \langle P_{\lambda(A \cap B)}e'_m, e'_n \rangle,$$

which is zero by the construction of e'_n, since $P_{\lambda(A \cap B)}e'_m \in E_m$.

Next, we claim that $E_n = (E_n \cap \lambda(A)) + (E_n \cap \lambda(A^c))$ for all n and all measurable $A \subseteq X$. This is because for any measurable $B \subseteq X$ we can write $P_{\lambda(B)}e'_n = P_{\lambda(A \cap B)}e'_n + P_{\lambda(B \setminus A)}e'_n$, so that every vector of the form $P_{\lambda(B)}e'_n$ belongs to $(E_n \cap \lambda(A)) + (E_n \cap \lambda(A^c))$. Since the span of these vectors is dense in E_n, the claim follows.

We can now use Lemma 5.23 to conclude that $\lambda_n(A) = E_n \cap \lambda(A)$ is an E_n-valued spectral measure on X for each n. Moreover, the natural Hilbert isomorphism between H and $\bigoplus E_n$ implements an equivalence between λ and the direct sum of the λ_n. Since each λ_n has a cyclic vector e'_n, the theorem now follows from Lemmas 5.21 and 5.22.   □

5.3   Operators

A linear map from a normed vector space E to itself is usually called an *operator*. We denote the space $\mathcal{B}(E, E)$ of bounded operators on E by $\mathcal{B}(E)$.

We are chiefly interested in operators on Hilbert spaces. There are several reasons for this. One of the more conceptual ones is their relevance to physics; in quantum mechanics any physical system is modeled by a Hilbert space, with any possible state of the system being represented by a unit vector in the space and with any transformation of the system — the effect of some action that might be performed on the system, such as rotating it, or letting it evolve for some amount of time — being represented by an operator acting on the space. Not every operator can play this role; it has to be unitary, a condition we will explain below.

Any observable quantity is supposed to be modeled by a different kind of operator, a self-adjoint operator. The intuition for using operators here is less clear because the result of an observation ought to be a measured value, not another physical state. We really need spectral theory to understand what operators have to do with observations.

Much is known about Hilbert space operators, although basic questions remain unanswered. For example, does every bounded operator on a separable Hilbert space have a nontrivial closed invariant subspace? A closed subspace E of H is *invariant* for $T \in \mathcal{B}(H)$ if $T(E) \subseteq E$; $\{0\}$ and H are always closed invariant subspaces, and the question is whether there must necessarily be any others. This is known as the invariant subspace problem.

In the remaining sections of this chapter we are going to study operators on a single Hilbert space. Thus, *we assume that H denotes a fixed separable Hilbert space for the rest of this book.*

What distinguishes $\mathcal{B}(H)$ from $\mathcal{B}(E)$ for other Banach spaces is the existence of adjoints. If E and F are normed linear spaces and $T \in \mathcal{B}(E, F)$, then there is a natural map $T' \in \mathcal{B}(F', E')$ defined by $T'f = f \circ T$ for $f \in F'$ (see Corollary 4.9). In particular, if $T \in \mathcal{B}(H)$ then $T' \in \mathcal{B}(H')$. The new feature in the Hilbert space setting is that we have a natural bijection between H and H', so that we can convert T' into an operator T^* in $\mathcal{B}(H)$.

Definition 5.25. Let $T \in \mathcal{B}(H)$ and define a map $h : H \to H'$ by setting $(hw)(v) = \langle v, w \rangle$; this map is a surjective isometry by Theorem 5.13. Then define the *adjoint T^** of T by $T^* = h^{-1} \circ T' \circ h$.

The map $h : H \to H'$ is linear in the real case, but in the complex case it is *antilinear*: it satisfies $h(v + w) = hv + hw$ and $h(av) = \bar{a}hv$. Thus $T' \circ h$ is also antilinear and $h^{-1} \circ T' \circ h$ is linear again.

Proposition 5.26. *Let $T \in \mathcal{B}(H)$. Then $T^* \in \mathcal{B}(H)$ and $\langle Tv, w \rangle = \langle v, T^*w \rangle$ for all $v, w \in H$.*

Proof. We just observed that T^* is linear. It is bounded because h and h^{-1} are isometries and T', being composition with T, satisfies $\|T'\| \leq \|T\|$.

Let $w \in H$. Then hw is the linear functional $v \mapsto \langle v, w \rangle$. So $T'(hw)$ is the linear functional $v \mapsto \langle Tv, w \rangle$, and $T^*w = h^{-1}(T'(hw))$ is the vector w' which satisfies $\langle v, w' \rangle = \langle Tv, w \rangle$ for all v. This proves the adjoint formula $\langle v, T^*w \rangle = \langle Tv, w \rangle$. $\square$

Note that the formula $\langle Tv, w \rangle = \langle v, T^*w \rangle$ characterizes T^*w. If we know what the inner product of T^*w is with every vector v, then we know what T^*w is.

If H has an orthonormal basis (e_n) then we can represent T by the matrix of scalars $\langle Te_m, e_n \rangle$. By linearity these values determine $\langle Tv, w \rangle$ for any v and w in the span of the basis, and then by continuity they determine all values of $\langle Tv, w \rangle$. The matrix of the adjoint has entries $\langle T^*e_m, e_n \rangle = \langle e_m, Te_n \rangle = \overline{\langle Te_n, e_m \rangle}$. Thus, it is the conjugate transpose of the matrix of T: we interchange m and n (i.e., we transpose the matrix) and then conjugate its entries. However, in infinite dimensions the matrix picture has limited value because there is no good way of diagnosing which infinite arrays of scalars correspond to bounded operators.

Proposition 5.27. *Let $S, T \in \mathcal{B}(H)$ and let $a, b \in \mathbf{F}$. Then*

 (i) $(aS + bT)^* = \bar{a}S^* + \bar{b}T^*$;
 (ii) $(ST)^* = T^*S^*$;
 (iii) $T^{**} = T$;
 (iv) $\|T^*\| = \|T\|$;
 (v) $\|T^*T\| = \|T\|^2$.

Proof.   The first two parts follow directly from the definition of the adjoint. We have

$$h^{-1} \circ (aS' + bT') \circ h = \bar{a}h^{-1} \circ S' \circ h + \bar{b}h^{-1} \circ T' \circ h$$

since h^{-1} is antilinear, and

$$h^{-1} \circ (ST)' \circ h = h^{-1} \circ T'S' \circ h = (h^{-1} \circ T' \circ h)(h^{-1} \circ S' \circ h).$$

Next, applying condition (ii) of Definition 5.1 to the adjoint formula yields $\langle w, Sv \rangle = \langle S^*w, v \rangle$, so that (taking $S = T^*$ and switching v and w)

$$\langle Tv, w \rangle = \langle v, T^*w \rangle = \langle T^{**}v, w \rangle$$

for all v and w, and hence $T = T^{**}$.

We have $\|T^*\| = \|T'\| \leq \|T\|$ since h and h^{-1} are isometries. But $\|T^{**}\| \leq \|T^*\|$ for the same reason, and since $T = T^{**}$ we conclude that $\|T\| = \|T^*\|$. Finally, $\|T^*T\| \leq \|T\|\|T^*\| = \|T\|^2$ follows from the preceding fact, while combining the equation

$$\langle T^*Tv, v \rangle = \langle Tv, Tv \rangle = \|Tv\|^2$$

with the Cauchy-Schwarz inequality $|\langle T^*Tv, v \rangle| \leq \|T^*T\|\|v\|^2$ and taking the supremum over $\|v\| = 1$ yields $\|T^*T\| \geq \|T\|^2$. $\square$

We are interested in operators that have special properties relative to their adjoints. Let $I \in \mathcal{B}(H)$ denote the identity operator defined by setting $Iv = v$ for all v.

Definition 5.28. Let $T \in \mathcal{B}(H)$. Then T is *self-adjoint* if $T = T^*$, *unitary* if $T^*T = TT^* = I$, and a *projection* if $T = T^* = T^2$.

We will consider these categories of operators in reverse order. Projections and unitaries have simple geometric interpretations, while analyzing self-adjoint operators will take us through the next two sections.

Recall that if E is a closed subspace of H then $P_E v$ denotes the projection onto E.

Proposition 5.29. *For any closed subspace E of H the map P_E is a projection. Every projection is the projection onto its range.*

Proof. Let E be a closed subspace of H. We must check that P_E is linear. Let $v, w \in H$ and write $v = v_1 + v_2$ and $w = w_1 + w_2$ with $v_1, w_1 \in E$ and $v_2, w_2 \in E^\perp$. Then $v_1 + v_2 \in E$ and $v_2 + w_2 \in E^\perp$, so that the expression $v + w = (v_1 + w_1) + (v_2 + w_2)$ verifies that $P_E(v + w) = v_1 + w_1 = P_E v + P_E w$. Likewise, $av = av_1 + av_2$ decomposes av as a vector in E plus a vector in $E^\perp$, and this shows that $P_E(av) = av_1 = aP_E v$. So P_E is linear. It is bounded because $\|v\|^2 = \|v_1\|^2 + \|v_2\|^2$, and hence $\|v_1\| \le \|v\|$. We have $P_E^2 = P_E$ because $P_E v \in E$, which implies that $P_E(P_E v) = P_E v$, for all $v \in H$. Finally, decomposing arbitrary v and w as above lets us calculate

$$\langle P_E v, w \rangle = \langle v_1, w_1 + w_2 \rangle = \langle v_1, w_1 \rangle = \langle v_1 + v_2, w_1 \rangle = \langle v, P_E w \rangle,$$

showing that P_E is self-adjoint. We conclude that P_E is a projection.

Now let $P \in \mathcal{B}(H)$ be any projection and let $E = P(H)$. It is clear that E is a subspace of H; to see that it is closed, let (v_n) be a Cauchy sequence in E and say $v_n \to v$. Then $P v_n = v_n$ for all n by the property $P = P^2$, so that $P v = v$ by continuity. Thus we have shown that the limit of any Cauchy sequence in E also belongs to E.

We know that $Pv = v = P_E v$ for all $v \in E$. For $v \in E^\perp$ we have $P_E v = 0$, and also $\langle Pv, w \rangle = \langle v, Pw \rangle = 0$ for all $w \in H$, which implies that $Pv = 0$. So P and P_E agree on both E and $E^\perp$, and we conclude that they must be equal. $\square$

This completely characterizes the structure of projections in $\mathcal{B}(H)$. For unitaries, we have an equally simple geometric characterization.

Proposition 5.30. *An operator $U \in \mathcal{B}(H)$ is unitary iff it is a Hilbert isomorphism of H with itself.*

Proof. Suppose U is unitary. Then $\langle Uv, Uw \rangle = \langle U^* U v, w \rangle = \langle v, w \rangle$ for all v and w in H, so U is a Hilbert isomorphism between H and $U(H)$. But the condition $UU^* = I$ forces $U(H)$ to equal H.

Conversely, suppose $U : H \cong H$ is a Hilbert isomorphism. Then we have $\langle v, w \rangle = \langle Uv, Uw \rangle = \langle U^* U v, w \rangle$ for all $v, w \in H$. This shows that $U^* U v = v$ for all v, i.e., $U^* U = I$. We also know that U is invertible, so U^{-1} must equal U^*, and therefore $U^* U = I$ holds as well. $\square$

We should note that the condition $U^*U = I$ alone implies that U is a Hilbert isomorphism between H and $U(H)$, but it does not ensure that U is surjective. For example, define $U \in \mathcal{B}(l^2)$ to be the map that takes the sequence $(a_1, a_2, a_3, \dots)$ to the sequence $(0, a_1, a_2, \dots)$. This is called the *unilateral shift*, and it is an isometry but not a unitary.

5.4 The continuous functional calculus

H continues to be a fixed separable Hilbert space. In this section and the next, we also fix a self-adjoint operator $T \in \mathcal{B}(H)$.

Let (X, μ) be a σ-finite measure space, let $\mathcal{H}$ be a measurable Hilbert bundle over X, and let $f \in L^\infty(X, \mu)$. For $\eta \in L^2(X, \mu; \mathcal{H})$ define the product $(f \cdot \eta)(x) = f(x)\eta(x)$ pointwise. Then the map $M_f : \eta \mapsto f \cdot \eta$ is clearly linear, and the computation

$$\|f \cdot \eta\|^2 = \int \|f(x) \cdot \eta(x)\|^2 \le \|f\|_\infty^2 \int \|\eta(x)\|^2 = \|f\|_\infty^2 \|\eta\|^2$$

shows that it is bounded, with $\|M_f\| \le \|f\|_\infty$. Conversely, choosing η to be supported on a region where $|f(x)| \ge \|f\|_\infty - \epsilon$ (using σ-finiteness to find such a region that has nonzero but finite measure) and taking $\epsilon \to 0$ yields that $\|M_f\| = \|f\|_\infty$. We call $M_f \in \mathcal{B}(L^2(X, \mu; \mathcal{H}))$ a *multiplication operator*.

The adjoint of M_f is easily computed. We have

$$\langle M_f \eta, \xi \rangle = \int \langle f(x)\eta(x), \xi(x) \rangle = \int \langle \eta(x), \overline{f(x)}\xi(x) \rangle = \langle \eta, M_{\bar{f}}\xi \rangle;$$

thus $M_f^* = M_{\bar{f}}$ is multiplication by the pointwise complex conjugate of f. In particular, M_f is self-adjoint if and only if f is almost everywhere real-valued. The spectral theorem for bounded self-adjoint operators will provide a converse assertion: every bounded self-adjoint operator can be realized as a multiplication operator on some $L^2(X, \mu; \mathcal{H})$. This result gives us a very good understanding of self-adjoint operators. Since multiplication operators on l_n^2 correspond to diagonal matrices in the standard basis, the spectral theorem can be seen as a far-reaching generalization of the finite-dimensional fact that Hermitian matrices are diagonalizable.

Let us pause to record a simple special case of the preceding discussion.

Proposition 5.31. *Let $E_1, \dots, E_n$ be orthogonal nonzero closed subspaces and let $a_1, \dots, a_n \in \mathbf{F}$. Then the norm of the operator $\sum a_i P_{E_i}$ is $\max |a_i|$.*

We can assume that $E_1 + \cdots + E_n = H$; if not, include one more projection to make up the deficit and assign it the coefficient 0. The proposition then follows from the above by identifying H with the L^2 sections of a bundle over the set $\{1, \ldots, n\}$ with fibers E_n. In this Hilbert bundle picture the operator $\sum a_i P_{E_i}$ is just multiplication by the vector $(a_1, \ldots, a_n)$, whose sup norm is $\max |a_i|$.

We will prove the spectral theorem in the next section. The first tool we need is the spectrum of an operator, which will play the role of the space X in the $L^2(X, \mu; \mathcal{H})$ model for T.

Definition 5.32. Let $S \in \mathcal{B}(H)$. The *spectrum* of S is the set $\mathrm{sp}(S)$ of scalars $z \in \mathbf{F}$ with the property that $S - zI$ is not invertible in $\mathcal{B}(H)$.

Recall that I is the identity operator on H. This definition may seem a little obscure, but some examples will help us to build up an intuition for it.

Example 5.33. Let M_f be a multiplication operator on some $L^2(X, \mu; \mathcal{H})$, with (X, μ) a σ-finite measure space and $f \in L^\infty(X, \mu)$. Then $\mathrm{sp}(M_f) = \mathrm{ess\,ran}(f)$. To see this, first suppose z is not in the essential range of f. Then for some $\epsilon > 0$ we have $|f(x) - z| \geq \epsilon$ for almost every x. This implies that the quantity $\frac{1}{|f(x) - z|}$ is at most $\frac{1}{\epsilon}$ for almost every x; in particular, it is essentially bounded. So $g = \frac{1}{f - z \cdot 1_X}$ belongs to $L^\infty(X, \mu)$, and we have $M_f - zI = M_{f - z \cdot 1_X}$ and hence
$$(M_f - zI)M_g = M_g(M_f - zI) = M_{(f - z \cdot 1_X)g} = I.$$
This shows that $M_f - zI$ is invertible, and we conclude that $\mathrm{sp}(M_f)$ is contained in $\mathrm{ess\,ran}(f)$.

Conversely, if z belongs to the essential range of f then the set $f^{-1}(\mathrm{ball}_\epsilon(z))$ has nonzero measure for each $\epsilon > 0$. Any function $\eta \in L^2(X, \mu; \mathcal{H})$ that is supported on this set will then satisfy $\|M_{f - z \cdot 1_X}\eta\| \leq \epsilon \|\eta\|$. This implies that the inverse of $M_{f - z \cdot 1_X} = M_f - zI$, were it to exist, would have norm at least $\frac{1}{\epsilon}$. As ϵ was arbitrary, we conclude that $M_f - zI$ is not invertible in $\mathcal{B}(H)$. Thus $\mathrm{ess\,ran}(f)$ is contained in $\mathrm{sp}(M_f)$.

We can also neatly characterize the spectrum if H is finite-dimensional, when an operator is invertible if and only if it has zero kernel.

Example 5.34. Suppose H is finite-dimensional and let $S \in \mathcal{B}(H)$. Then $S - zI$ fails to be invertible if and only if there exists a nonzero vector $v \in H$ such that $(S - zI)v = 0$. Rewriting this condition as $Sv = zv$ shows that $z \in \mathrm{sp}(S)$ if and only if z is an eigenvalue of S.

This example reveals that the spectrum can be empty, at least if $\mathbf{F} = \mathbf{R}$. The rotation matrix $\begin{bmatrix} \cos\theta & -\sin\theta \\ \sin\theta & \cos\theta \end{bmatrix}$ has no real eigenvalues unless θ is a multiple of π. Even if the scalars are complex the matrix $\begin{bmatrix} 0 & 1 \\ 0 & 0 \end{bmatrix}$ shows that the spectrum might consist only of the single point 0. However, for self-adjoint operators the spectrum has to be more substantial. We will now show that the norm of T can always be recovered from its spectrum. (Recall that we are assuming T is self-adjoint.)

This result uses three elementary facts. The first is the identity

$$\|v + w\|^2 + \|v - w\|^2 = 2\|v\|^2 + 2\|w\|^2$$

for all $v, w \in H$. This is called the *parallelogram law* because the vectors $v \pm w$ describe the diagonals of a parallelogram with sides v and w. It is proven by expanding the squared norms on the left as inner products. The cross terms cancel.

The second is the *real polarization identity*

$$\Re\langle Sv, w\rangle = \frac{1}{4}\left(\langle S(v+w), v+w\rangle - \langle S(v-w), v-w\rangle\right)$$

for $v, w \in H$ and $S \in \mathcal{B}(H)$ self-adjoint. It is also a simple computation: after expanding the inner products on the right side and subtracting, only the cross terms survive. We use self-adjointness to rewrite $\langle Sw, v\rangle$ as $\overline{\langle Sv, w\rangle}$.

The last fact we need states that if $S \in (\mathcal{B}(H))_1$ then $I - S$ is invertible in $\mathcal{B}(H)$. To see this let $r = \|S\| < 1$. Then inductively $\|S^n\| \leq r^n$ for all n, which means that the series $\sum_{n=0}^{\infty} S^n$ is absolutely summable in $\mathcal{B}(H)$. (We use the convention that $S^0 = I$.) We have

$$(I - S)\sum_{n=0}^{\infty} S^n = \lim_{n \to \infty}(I - S)\sum_{k=0}^{n} S^k = \lim_{n \to \infty}(I - S^{n+1}) = I$$

since the sum telescopes, and $(\sum S^n)(I - S) = I$ similarly. Thus $I - S$ is invertible. The series $\sum_{n=0}^{\infty} S^n$ is called a *Neumann series*.

Proposition 5.35. $\|T\| = \sup\{|z| : z \in \mathrm{sp}(T)\}$.

Proof. Let $A = \sup\{|\langle Tv, v\rangle| : \|v\| = 1\}$ and $B = \sup\{|z| : z \in \mathrm{sp}(T)\}$. We show that $\|T\| \leq A \leq B \leq \|T\|$.

First, by scaling we have $|\langle Tv, v\rangle| \leq A\|v\|^2$ for all $v \in H$. So if $\|v\| = \|w\| = 1$ then we can use real polarization and the parallelogram law to estimate

$$|\Re\langle Tv, w\rangle| = \frac{1}{4}|\langle T(v+w), v+w\rangle - \langle T(v-w), v-w\rangle|$$

$$\leq \frac{A}{4}(\|v + w\|^2 + \|v - w\|^2)$$
$$= \frac{A}{2}(\|v\|^2 + \|w\|^2) = A.$$

Replacing v with $v' = \frac{|\langle Tv, w \rangle|}{\langle Tv, w \rangle}v$ in this inequality yields

$$A \geq |\Re\langle Tv', w \rangle| = |\langle Tv, w \rangle|.$$

As this holds for all unit vectors v and w, we conclude that $\|T\| \leq A$.

For the second inequality, find a sequence of unit vectors v_n such that $|\langle Tv_n, v_n \rangle| \to A$. Since T is self-adjoint we have $\langle Tv, v \rangle = \langle v, Tv \rangle = \overline{\langle Tv, v \rangle}$, so that $\langle Tv, v \rangle$ is real for any vector v. Thus, by passing to a subsequence we can assume that $\langle Tv_n, v_n \rangle$ converges to either A or $-A$. Let $z = \lim\langle Tv_n, v_n \rangle = \pm A$. Then (using $\|T\| \leq A$, which we just proved) we have

$$\|Tv_n - zv_n\|^2 = \|Tv_n\|^2 - 2z\langle Tv_n, v_n \rangle + z^2\|v_n\|^2$$
$$\leq A^2 - 2z\langle Tv_n, v_n \rangle + A^2 \to 0$$

as $n \to \infty$. That is, $(T - zI)v_n \to 0$ as $n \to \infty$, and this shows that $T - zI$ cannot have a bounded inverse. So $z = \pm A$ belongs to sp(T), and therefore $A \leq B$.

Finally, to verify that $B \leq \|T\|$, let $z > \|T\|$. Then $T - zI = z(\frac{1}{z}T - I)$, and by Neumann series $\frac{1}{z}T - I$ is invertible since $\|\frac{1}{z}T\| < 1$, so $T - zI$ must be invertible. Thus, any element of sp(T) must have modulus at most $\|T\|$, i.e., $B \leq \|T\|$. $\qquad\square$

This is a valuable result because it relates the operator norm of T to an object, the spectrum of T, which is defined in purely algebraic terms.

Corollary 5.36. *The spectrum of T is a nonempty compact subset of* **R**.

Proof. It immediately follows from Proposition 5.35 both that sp(T) cannot be empty and that it is a bounded subset of **F**. To see that it is closed, suppose $z \notin$ sp(T) and let $S = (T - zI)^{-1}$; then for any $w \in$ ball$_{1/\|S\|}(z)$ we have

$$\|(T - wI)S - I\| = \|((T - wI) - (T - zI))S\|$$
$$= \|(z - w)S\| < 1,$$

which implies that $(T - wI)S$ is invertible by Neumann series. Thus $T - wI$ must be right invertible, and a similar calculation shows that it is left invertible. Thus the complement of sp(T) contains a ball about each of its points. We conclude that sp(T) is closed.

We still have to prove that $\mathrm{sp}(T) \subset \mathbf{R}$ in the complex setting. Let $z \in \mathbf{C} \setminus \mathbf{R}$, so that $z = a + bi$ with $b \neq 0$. We must show that $T - (a + bi)I$ is invertible. Rewriting $T - (a + bi)I$ as $b(\frac{1}{b}(T - aI) - iI)$ and observing that $\frac{1}{b}(T - aI)$ is also self-adjoint allows us to reduce to the case where $z = i$. Now we have

$$\begin{aligned}
\|(T - iI)v\|^2 &= \langle (T - iI)v, (T - iI)v \rangle \\
&= \langle (T + iI)(T - iI)v, v \rangle \\
&= \langle (T^2 + I)v, v \rangle \\
&= \|Tv\|^2 + \|v\|^2
\end{aligned}$$

for any $v \in H$. Thus $\|v\| \leq \|(T - iI)v\| \leq \|T - iI\|\|v\|$ for all v, and hence $T - iI$ is a Banach space isomorphism between H and its image. It follows that this image must be a closed subspace of H, so all we need to do to establish invertibility is to show that $T - iI$ is surjective. By Lemma 5.12 it will suffice to show that no nonzero vector in H is orthogonal to $(T - iI)(H)$.

Thus let $w \in H$ and suppose $\langle (T - iI)v, w \rangle = 0$ for all $v \in H$. Taking $v = (T + iI)w$ then yields

$$0 = \langle (T - iI)(T + iI)w, w \rangle = \langle (T^2 + I)w, w \rangle = \|Tw\|^2 + \|w\|^2,$$

which indeed implies that $w = 0$. We conclude that $T - iI$ is invertible. $\square$

The real version of the fact that $T - zI$ is invertible for all $z \in \mathbf{C} \setminus \mathbf{R}$ says that $T^2 + a^2 I$ is invertible for all nonzero $a \in \mathbf{R}$. This is proven in a similar way: first calculate

$$\begin{aligned}
\|(T^2 + a^2 I)v\|^2 &= \langle (T^2 + a^2 I)v, (T^2 + a^2 I)v \rangle \\
&= \langle (T^4 + 2a^2 T^2 + a^4 I)v, v \rangle \\
&= \|T^2 v\|^2 + 2a^2 \|Tv\|^2 + a^4 \|v\|^2,
\end{aligned}$$

which shows that $a^2 \|v\| \leq \|(T^2 + a^2 I)v\|$ for all v, and then establish surjectivity by assuming $\langle (T^2 + a^2 I)v, w \rangle = 0$ for all v and inferring $\|Tw\|^2 + a^2 \|w\|^2 = 0$, and hence $w = 0$, by setting $v = w$.

We now have almost all of the technical tools we need to achieve the main goal of this section, constructing the continuous functional calculus. The last piece is the following proposition, which is known as the spectral mapping theorem.

In order to state this result, we need the following notation. If $p(x) = \sum_{k=0}^{n} a_k x^k$ is a polynomial in x with coefficients in $\mathbf{F}$, let $p(T)$ be the operator $p(T) = \sum_{k=0}^{n} a_k T^k$, using the convention that $T^0 = I$.

Proposition 5.37. *Let p be a polynomial. Then $\mathrm{sp}(p(T)) = p(\mathrm{sp}(T))$.*

Proof. The proof is slightly different in the real and complex cases. Consider the complex case first. For any $z \in \mathbf{C}$, factor $p(x) - z$ as

$$p(x) - z = b(x - b_1) \cdots (x - b_n),$$

where n is the degree of p and the values b_i are the roots of the equation $p(x) = z$. We then have

$$p(T) - zI = b(T - b_1I) \cdots (T - b_nI).$$

Since the factors on the right commute, we can conclude that $p(T) - zI$ fails to be invertible if and only if $T - b_iI$ fails to be invertible for some i. That is, $z \in \mathrm{sp}(p(T))$ if and only if $w \in \mathrm{sp}(T)$ for some w satisfying $p(w) = z$. This shows that $\mathrm{sp}(p(T)) = p(\mathrm{sp}(T))$.

In the real case, we assume that $z \in \mathbf{R}$ and $p(x)$ has real coefficients. We might not be able to factor $p(x) - z$ into linear factors, but we can factor it into linear factors $x - b_i$ and irreducible quadratic factors $(x - a_j)^2 + c_j^2$ with $c_j \neq 0$. We know from the comment preceding this theorem that $(T - a_jI)^2 + c_j^2I$ is always invertible. So the quadratic factors are irrelevant to the invertibility comparison between $p(T) - zI$ and $(T - b_1I) \cdots (T - b_kI)$, and we conclude as before that $p(T) - z$ fails to be invertible if and only if $T - b_iI$ fails to be invertible for some real root b_i of the equation $p(x) = z$. Thus again $\mathrm{sp}(p(T)) = p(\mathrm{sp}(T))$. $\square$

Let X be a locally compact Hausdorff space. A *homomorphism* from $C_0(X)$ to $\mathcal{B}(H)$ is a linear map $\Phi : C_0(X) \to \mathcal{B}(H)$ which satisfies the law $\Phi(fg) = \Phi(f)\Phi(g)$ for all $f, g \in C_0(X)$. If it also satisfies $\Phi(\bar{f}) = \Phi(f)^*$ for all $f \in C_0(X)$ then it is *self-adjoint*. It is *unital* if X is compact and $\Phi(1_X) = I$.

The *continuous functional calculus* is the name for the homomorphism constructed in the following theorem.

Theorem 5.38. *(Continuous functional calculus) Let T be a bounded self-adjoint operator on a separable Hilbert space H. Then the map which takes the polynomial p to the operator $p(T)$ extends to an isometric, unital, self-adjoint homomorphism $\Phi : C(\mathrm{sp}(T)) \to \mathcal{B}(H)$.*

Proof. Suppose first that p is a real polynomial. Then $p(T)$ is self-adjoint, and it follows from Propositions 5.35 and 5.37 that

$$\|p(T)\| = \sup\{|p(t)| : t \in \mathrm{sp}(T)\} = \|p|_{\mathrm{sp}(T)}\|_\infty.$$

If $\mathbf{F} = \mathbf{C}$ and p has complex coefficients then we can apply the preceding to the real polynomial $p\bar{p} = |p|^2$ and obtain

$$\|p|_{\mathrm{sp}(T)}\|_\infty^2 = \|\bar{p}p|_{\mathrm{sp}(T)}\|_\infty = \|p(T)^*p(T)\| = \|p(T)\|^2.$$

Applying this to $p - q$ shows that if p and q agree on $\mathrm{sp}(T)$ then $p(T) = q(T)$, so the map $p|_{\mathrm{sp}(T)} \mapsto p(T)$ is well-defined, and moreover it is isometric. It is unital by the definition of $p(T)$. Since the polynomials are dense in $C(\mathrm{sp}(T))$, this map extends to an isometry Φ from $C(\mathrm{sp}(T))$ into $\mathcal{B}(H)$. The equations expressing that Φ is a self-adjoint homomorphism follow for all $f, g \in C(\mathrm{sp}(T))$ by continuity from their validity for polynomials. $\quad\square$

In analogy to the notation $p(T)$, for any $f \in C(\mathrm{sp}(T))$ we write $f(T)$ for the operator $\Phi(f)$ constructed in this theorem.

5.5   The spectral theorem

H continues to be a fixed separable Hilbert space and T continues to be a fixed bounded, self-adjoint operator on H.

We are going to use our analysis of spectral measures from Section 5.2 to find a measurable Hilbert bundle such that H is Hilbert isomorphic to $L^2(X, \mu; \mathcal{H})$ in a way that makes T correspond to a multiplication operator. This spectral measure will be constructed using not the continuous functional calculus of the last section, but an extension of it called the Borel functional calculus.

The idea is this. We can form polynomials in T just using the algebraic operations in $\mathcal{B}(H)$. In constructing the continuous functional calculus we take norm limits to pass from polynomials to continuous functions on $\mathrm{sp}(T)$. The Borel functional calculus goes one step further, using weak* limits to pass from continuous functions to bounded measurable functions.

$\mathcal{B}(H)$ is a dual space by the comment following Corollary 4.21. Explicitly, if (e_n) is an orthonormal basis of H then the maps $\omega_{m,n} : S \mapsto \langle Se_m, e_n \rangle$ form a countable family of bounded linear functionals on $\mathcal{B}(H)$ which satisfy the hypotheses of Corollary 4.21. So by that result $\mathcal{B}(H) \cong E'$ where $E = \mathrm{span}\{\omega_{m,n}\} \subseteq \mathcal{B}(H)'$. This provides us with a weak* topology on $\mathcal{B}(H)$, but it depends on a choice of basis. The canonical predual of $\mathcal{B}(H)$ is $\overline{E}$, the closure of E in $\mathcal{B}(H)'$. This closure is usually denoted $\mathcal{C}_1(H)$ for reasons we need not go into. It contains every linear functional of the form $S \mapsto \langle Sv, w \rangle$ for $v, w \in H$ and it does not depend on a choice of basis. However, recall from Theorem 4.3 that replacing E with its closure does not change the weak* topology on bounded sets.

We use the same terminology (homomorphism, unital, self-adjoint) for maps from $L^\infty(X, \mu)$ into $\mathcal{B}(H)$ as for maps from $C(X)$ into $\mathcal{B}(H)$. We now show how to extend homomorphisms from $C(X)$ into $\mathcal{B}(H)$ to homomorphisms from $L^\infty(X, \mu)$ into $\mathcal{B}(H)$. We state this result in a form which applies both to the current setting and to more general settings that we will consider in the next section.

Say that a positive Borel measure is *strictly positive* if every nonempty open set has positive measure. This condition is needed to ensure that the essential sup norm of any continuous function equals its sup norm. (If there were a nonempty open set with measure zero then a continuous function might achieve its supremum on that set, but it would not count toward its essential sup.)

Lemma 5.39. *Let X be a compact metrizable space and let $\Phi : C(X) \to \mathcal{B}(H)$ be an isometric, unital, self-adjoint homomorphism. Then there is a finite, strictly positive Borel measure μ on X and an isometric, weak* homeomorphic, self-adjoint homomorphism $\tilde{\Phi} : L^\infty(X, \mu) \to \mathcal{B}(H)$ which extends Φ.*

Proof. Let (e_n) be an orthonormal basis of H and let E be the span of the corresponding bounded linear functionals $\omega_{m,n}$, as discussed above. Define $\Phi' : E \to C(X)' \cong M(X)$ by $\Phi'(\omega) = \omega \circ \Phi$. (This is the restriction to E of the dual map described in Corollary 4.9.) Thus if $\mu_{m,n} = \Phi'(\omega_{m,n})$ then we have

$$\int f \, d\mu_{m,n} = \langle \Phi(f) e_m, e_n \rangle$$

for all $f \in C(X)$.

Since $\|\omega_{m,n}\| = 1$ for all m, n, it follows that $\|\mu_{m,n}\| \leq 1$ for all m, n. So define $\mu = \sum_{m,n} \frac{1}{2^{m+n}} |\mu_{m,n}|$. The measure μ is strictly positive because if it were not we would have $\int f \, d\mu = 0$ for some nonzero positive continuous function f, which implies that $\int f \, d\mu_{m,n} = 0$ for all m and n, and hence that $\langle \Phi(f) e_m, e_n \rangle = 0$ for all m and n. But this can only happen if $\Phi(f) = 0$, which contradicts the assumption that Φ is isometric.

We have $\mu_{m,n} \ll \mu$ for all m and n; by linearity, the measure $\Phi'(\omega)$ is absolutely continuous with respect to μ for every $\omega \in E$. Thus, for each $\omega \in E$ let g_ω be the Radon-Nikodym derivative of $\Phi'(\omega)$ with respect to μ. The map $\omega \mapsto g_\omega$ is then a nonexpansive linear map from E into $L^1(X, \mu)$ because $|\int f g_\omega \, d\mu| = |\omega(\Phi(f))| \leq \|\omega\| \|f\|_\infty$ for all $f \in C(X)$, and therefore $\|g_\omega\|_1 \leq \|\omega\|$ by Proposition 4.46. So the dual map (Corollary 4.9) is a weak* continuous linear map $\tilde{\Phi}$ from $L^\infty(X, \mu) \cong L^1(X, \mu)'$ into $\mathcal{B}(H) \cong E'$. By Corollary 4.14 it is also weak* continuous into $\mathcal{B}(H)$ with the weak* topology coming from its standard predual $\mathcal{C}_1(H)$. The map $\tilde{\Phi}$

is characterized by the property

$$(\tilde{\Phi}(f))(\omega) = \int f g_\omega \, d\mu$$

for all $f \in L^\infty(X, \mu)$ and $\omega \in E$. Thus, for any $f \in C(X)$ and all m, n we have

$$\langle \tilde{\Phi}(f) e_m, e_n \rangle = \int f \, d\mu_{m,n} = \langle \Phi(f) e_m, e_n \rangle,$$

so that $\tilde{\Phi}(f) = \Phi(f)$.

To prove that $\tilde{\Phi}$ is a self-adjoint homomorphism, choose $f, g \in L^\infty(X, \mu)$ and use Proposition 4.46 to find bounded sequences $(f_n), (g_n) \subset C(X)$ which converge weak* to f and g. Then it is easy to see that $\bar{f}_n$ converges weak* to $\bar{f}$, so weak* continuity of $\tilde{\Phi}$ implies that

$$\tilde{\Phi}(\bar{f}) = \lim \Phi(\bar{f}_n) = \lim \Phi(f_n)^* = \tilde{\Phi}(f)^*.$$

This shows that $\tilde{\Phi}$ is self-adjoint. It is a homomorphism because $f_m g_n \to f g_n$ weak* for all n, so that

$$\tilde{\Phi}(f g_n) = \lim_{m \to \infty} \Phi(f_m g_n) = \lim_{m \to \infty} \Phi(f_m) \Phi(g_n) = \tilde{\Phi}(f) \Phi(g_n)$$

for all n, and then $f g_n \to f g$ weak*, so that

$$\tilde{\Phi}(f g) = \lim_{n \to \infty} \tilde{\Phi}(f g_n) = \lim_{n \to \infty} \tilde{\Phi}(f) \Phi(g_n) = \tilde{\Phi}(f) \tilde{\Phi}(g).$$

This completes the proof that $\tilde{\Phi}$ is a self-adjoint homomorphism.

Next, we check that $\tilde{\Phi}$ is isometric. It is enough to show that $\|h\|_\infty = \|\tilde{\Phi}(h)\|$ for simple functions h, since these functions are norm dense in $L^\infty(X, \mu)$. The key observation is that $\mu(A) > 0$ implies $\tilde{\Phi}(1_A) \neq 0$. This is because if $\mu(A) > 0$ then for some m, n we must have $|\mu_{m,n}|(A) > 0$, and hence $\mu_{m,n}(B) \neq 0$ for some $B \subseteq A$; thus $\langle \tilde{\Phi}(1_B) e_m, e_n \rangle = \int 1_B \, d\mu_{m,n} \neq 0$, and since $\tilde{\Phi}(1_A) \tilde{\Phi}(1_B) = \tilde{\Phi}(1_B)$ we conclude that $\tilde{\Phi}(1_A) \neq 0$. Now let $\sum a_i 1_{A_i}$ be the standard form of a simple function h, with $\mu(A_i) > 0$ for all i. Since $1_{A_i} = \bar{1}_{A_i} = 1_{A_i}^2$, the operators $\tilde{\Phi}(1_{A_i})$ have the same properties, i.e., they are projections. Moreover, $1_{A_i} 1_{A_j} = 0$ for $i \neq j$ implies that these projections have orthogonal ranges. Thus

$$\|\tilde{\Phi}(h)\| = \left\| \sum a_i \tilde{\Phi}(1_{A_i}) \right\| = \max |a_i| = \|h\|_\infty$$

by Proposition 5.31. We can now conclude that $\tilde{\Phi}$ is isometric.

It follows from Corollary 1.41 that $\tilde{\Phi}$ restricts to a weak* homeomorphism between $[L^\infty(X, \mu)]_1$ and its image, and the inverse map is then weak* continuous by Corollary 4.14. $\qquad\qquad\qquad\qquad\qquad\qquad\square$

The next result, the Borel functional calculus for self-adjoint operators, follows immediately from Lemma 5.39 and the properties of the continuous functional calculus given in Theorem 5.38.

Theorem 5.40. *(Borel functional calculus) Let T be a bounded self-adjoint operator on a separable Hilbert space H. Then there is a finite, strictly positive Borel measure μ on $\mathrm{sp}(T)$ such that the map $f \mapsto f(T)$ from $C(\mathrm{sp}(T))$ to $\mathcal{B}(H)$ extends to an isometric, weak* homeomorphic, unital, self-adjoint homomorphism from $L^\infty(\mathrm{sp}(T), \mu)$ into $\mathcal{B}(H)$.*

In the sequel will also use the notation $f(T)$ for $f \in L^\infty(\mathrm{sp}(T), \mu)$.

We are now ready to present the spectral theorem for bounded self-adjoint operators. We give it first in a more general form which will also be useful in the next section.

Lemma 5.41. *Let (X, μ) be a finite measure space and let $\Phi : L^\infty(X, \mu) \to \mathcal{B}(H)$ be an isometric, weak* continuous, unital, self-adjoint homomorphism. Then there is a measurable Hilbert bundle $\mathcal{H}$ over X and a Hilbert isomorphism $U : H \cong L^2(X, \mu; \mathcal{H})$ such that $\Phi(f) = U^{-1}M_f U$ for all $f \in L^\infty(X, \mu)$.*

Proof. As in the proof of Lemma 5.39, for any measurable $A \subseteq X$ the algebraic properties $1_A = 1_A^2 = \bar{1}_A$ imply that the operator $\Phi(1_A)$ is a projection. Let $\lambda(A)$ be the closed subspace of H such that $\Phi(1_A) = P_{\lambda(A)}$.

We claim that λ is an H-valued spectral measure on X. Condition (i) of Definition 5.19 is trivial. For condition (ii) just observe that $A \cap B = \emptyset$ implies $1_A \cdot 1_B = 0$. Thus $P_{\lambda(A)}P_{\lambda(B)} = 0$, which shows that $\lambda(A) \perp \lambda(B)$. Finally, let (A_n) be a sequence of disjoint measurable sets. Observe that $B \subseteq A$ implies that $1_A \cdot 1_B = 1_B$, and hence $P_{\lambda(A)}P_{\lambda(B)} = P_{\lambda(B)}$, and therefore $\lambda(B) \subseteq \lambda(A)$. Thus $\lambda(A_k) \subseteq \lambda(\bigcup A_n)$ for each k, and hence the closed span of the $\lambda(A_n)$ is contained in $\lambda(\bigcup A_n)$. For the reverse containment, use the fact that $1_{\bigcup_{n=1}^{N} A_n} \to 1_{\bigcup_{n=1}^{\infty} A_n}$ weak* in $L^\infty(X, \mu)$ (by DCT); thus $P_{\lambda(\bigcup_{n=1}^{N} A_n)} \to P_{\lambda(\bigcup_{n=1}^{\infty} A_n)}$ weak* in $\mathcal{B}(H)$, which means that the closed span of the $\lambda(A_n)$ must be all of $\lambda(\bigcup A_n)$. Otherwise there would exist a unit vector $v \in \lambda(\bigcup A_n)$ which is orthogonal to each $\lambda(A_n)$, and we would have $\langle P_{\lambda(\bigcup_{n=1}^{\infty} A_n)}v, v \rangle = 1$ but

$$\langle P_{\lambda(\bigcup_{n=1}^{N} A_n)}v, v \rangle = \langle (P_{\lambda(A_1)} + \cdots + P_{\lambda(A_N)})v, v \rangle = 0$$

for all N, which contradicts weak* convergence. We conclude that λ is a spectral measure.

Let ν, $\mathcal{H}$, and $U : H \cong L^2(X, \nu; \mathcal{H})$ be the measure, Hilbert bundle, and Hilbert isomorphism provided by Theorem 5.24. Then we have $U(\lambda(A)) = L^2(A, \nu|_A; \mathcal{H}|_A)$ for all measurable $A \subseteq X$. Thus

$$\Phi(1_A) = P_{\lambda(A)} = U^{-1} M_{1_A} U$$

for all measurable $A \subseteq \mathrm{sp}(T)$. Taking linear combinations yields $\Phi(h) = U^{-1} M_h U$ for all simple functions h on X, and taking norm limits then yields $\Phi(f) = U^{-1} M_f U$ for all $f \in L^\infty(X, \nu; \mathcal{H})$.

We complete the proof by showing that the measures μ and ν are mutually absolutely continuous. This will allow us to replace ν with μ in the above by using the weighting technique employed in the first part of the proof of Lemma 5.22. To verify this mutual absolute continuity, first observe that the fact that Φ is isometric tells us that $\mu(A) = 0$ if and only if $\Phi(1_A) = 0$, which is equivalent to the condition $\lambda(A) = \{0\}$. The measure ν, on the other hand, immediately satisfies $\nu(A) = 0$ if and only if $L^2(A, \nu|_A; \mathcal{H}) = \{0\}$. So $\mu(A) = 0$ if and only if $\nu(A) = 0$. $\square$

Combining this lemma with Theorem 5.40 and taking f to be the polynomial $f = x$ yields $T = U^{-1} M_x U$. Thus we have the following.

Theorem 5.42. *(Spectral theorem) Let T be a bounded self-adjoint operator on a separable Hilbert space H. Then there is a finite, strictly positive Borel measure μ on $\mathrm{sp}(T)$, a measurable Hilbert bundle $\mathcal{H}$ over $\mathrm{sp}(T)$, and a Hilbert isomorphism $U : H \cong L^2(\mathrm{sp}(T), \mu; \mathcal{H})$ such that $T = U^{-1} M_x U$.*

We mentioned earlier that self-adjoint operators play the role of observable quantities in quantum mechanics. The spectral measure λ constructed in the proof of Lemma 5.41 can help give some intuition for this association. The idea is that $\mathrm{sp}(T)$ represents the possible values of the measurement modeled by T. If the system is in a state that corresponds to a vector in $\lambda(A)$ then a measurement of the observable represented by T will yield a result in A. A general unit vector v can be decomposed into components $v_1 \in \lambda(A)$ and $v_2 \in \lambda(A^c)$, and $\|v_1\|^2$ and $\|v_2\|^2$ will respectively be the probabilities of obtaining a measured value in or out of A for a system in the state represented by v.

Another way to say this is in terms of the scalar-valued measure $A \mapsto \|P_{\lambda(A)} v\|^2$ used in Lemma 5.21. This will be a measure on $\mathrm{sp}(T)$ which, for each measurable $A \subseteq X$, gives the probability of obtaining a measured value in A.

5.6   Abelian operator algebras

The spectral theorem shows us how to realize a given self-adjoint operator as a multiplication operator. In this section, assuming complex scalars, we will present two grand generalizations of this result which allow us to simultaneously realize entire families of commuting self-adjoint operators in this form.

We already know something along these lines. Although the spectral theorem as we presented it was only formulated for a single operator T, the proof actually realized as a multiplication operator not just T, but every operator $f(T)$ in its Borel functional calculus. The entire algebra $L^\infty(\mathrm{sp}(T), \mu)$ naturally acts on $L^2(\mathrm{sp}(T), \mu; \mathcal{H})$ as multiplication operators.

When we are working with a family of commuting operators it is natural to assume that the family is an algebra (i.e., it is a linear space that is stable under products). There is no loss in generality in doing so because all of the operators in an algebra generated by a commuting family will commute with each other. We can also take norm limits without losing commutativity, or even weak* limits, as can be seen by using the two-step process for dealing with products that appeared in the proof of Lemma 5.39. Finally, if the original commuting family consists of self-adjoint operators then the algebra they generate will automatically be stable under adjoints, and this too will remain the case after taking the norm or weak* closure. All of this should help to motivate why the following definition describes good structures for doing spectral theory.

Definition 5.43. Assume $\mathbf{F} = \mathbf{C}$. A *C*-algebra* is a norm closed subspace of $\mathcal{B}(H)$ which is stable under adjoints and products. It is a *von Neumann algebra* if it is weak* closed. An algebra of either type is *abelian* if any two operators in it commute.

One of the basic operations we can perform within C*- and von Neumann algebras is the decomposition of an arbitrary operator T into its real and imaginary parts, $\Re T = \frac{1}{2}(T + T^*)$ and $\Im T = \frac{1}{2i}(T - T^*)$, which are both self-adjoint operators. It follows that every C*- and von Neumann algebra is spanned by its self-adjoint elements. This basic fact fails in the case of real scalars because we cannot form $\Im T$. So it is sensible to assume $\mathbf{F} = \mathbf{C}$ at this point. (But we note that Theorems 5.45 and 5.46 still hold when $\mathbf{F} = \mathbf{R}$ provided we assume the algebras are spanned by their self-adjoint elements.)

We have already seen some examples of abelian C*- and von Neumann algebras. When we constructed the continuous functional calculus (Theorem 5.38), we found that the set of operators $f(T)$ for $f \in C(\mathrm{sp}(T))$ was stable under sums, scalar products, operator products, and adjoints, in a way that just carried over the corresponding operations on $C(\mathrm{sp}(T))$. Moreover, since the continuous functional calculus is isometric, this set must be closed in norm. So it is a C*-algebra.

Likewise, the set of operators $f(T)$ for $f \in L^\infty(\mathrm{sp}(T), \mu)$ coming from the Borel functional calculus is a von Neumann algebra. Since the Borel functional calculus is an isometry, the unit ball of this set is the image of the unit ball of $L^\infty(\mathrm{sp}(T), \mu)$, and by weak* continuity it must be weak* compact and hence weak* closed. So the set itself is weak* closed by the Krein-Smulian theorem.

We can generalize the preceding examples as follows. Suppose X is any compact metrizable space, μ is a finite, strictly positive Borel measure on X, and $\mathcal{H}$ is a measurable Hilbert bundle over X. Then the map $f \mapsto M_f$ takes $C(X)$ and $L^\infty(X, \mu)$ isometrically into $\mathcal{B}(L^2(X, \mu; \mathcal{H}))$ and their images are respectively an abelian C*-algebra and an abelian von Neumann algebra. We will now show that, with minor qualification, this construction is general.

Lemma 5.44. *Let $T_1, \ldots, T_N \in (\mathcal{B}(H))_1$ be commuting self-adjoint operators. Then there is a nonexpansive, unital, self-adjoint homomorphism $\Phi : C([-1, 1])^N \to \mathcal{B}(H)$ taking the coordinate function x_n to the operator T_n for $1 \le n \le N$. The induced map from $C([-1, 1]^N)/\ker \Phi$ to $\mathcal{B}(H)$ is isometric.*

Proof. For $1 \le n \le N$ let μ_n be the measure on $\mathrm{sp}(T_n)$ provided by Theorem 5.40. We may regard μ_n as a measure on $[-1, 1)$ and the Borel functional calculus as a weak* continuous, unital, self-adjoint homomorphism from $L^\infty([-1, 1), \mu_n)$ into $\mathcal{B}(H)$ which takes the polynomial $p(x) = x$ to the operator T_n. Since T_m and T_n commute for all m and n, any two polynomials in the T_n's must commute, and then by continuity all operators in the ranges of their Borel functional calculi must commute.

Fix $k \in \mathbf{N}$ and partition the interval $[-1, 1)$ into 2^{k+1} subintervals $I_j = [\frac{j}{2^k}, \frac{j+1}{2^k})$ for $-2^k \le j < 2^k$. Consider simple functions $h = \sum a_i 1_{A_i}$ on the cube $[-1, 1)^N$ where the A_i are cubes of the form $I_{j_1} \times \cdots \times I_{j_N}$. We will call such h *dyadic simple functions*. Define $\Phi(h) \in \mathcal{B}(H)$ by setting

$$\Phi(1_{A_i}) = 1_{I_{j_1}}(T_1) \cdots 1_{I_{j_n}}(T_N)$$

for $A_i = I_{j_1} \times \cdots \times I_{j_N}$ and extending linearly. Note that the operators $\Phi(1_{A_i})$ are projections and their ranges are orthogonal. This defines a functional calculus for the dyadic simple functions on $[-1, 1)^N$ for a given value of k.

If we let k vary, the resulting functional calculi are compatible because

$$1_{[\frac{2j}{2^{k+1}}, \frac{2j+1}{2^{k+1}})}(T_i) + 1_{[\frac{2j+1}{2^{k+1}}, \frac{2j+2}{2^{k+1}})}(T_i) = 1_{[\frac{j}{2^k}, \frac{j+1}{2^k})}(T_i).$$

So we get a single functional calculus which defines an operator $\Phi(h)$ for every dyadic simple function h. Moreover, when we restrict to functions that depend on only one coordinate the result agrees with the Borel functional calculus for that T_n. We have $\|\Phi(h)\| \leq \max |a_i|$ by Proposition 5.31, with a possible inequality because we may have $\Phi(1_{A_i}) = 0$ for some cubes A_i, and these terms have to be removed from h before applying the proposition. Thus Φ is nonexpansive.

By continuity we can extend Φ to all functions on $[-1, 1)^N$ which are uniformly approximated by dyadic simple functions. This includes the restrictions of all functions in $C([-1, 1]^N)$ to $[-1, 1)^N$, so we get a map $\Phi : C([-1, 1]^N) \to \mathcal{B}(H)$, and it is a unital, self-adjoint homomorphism since the equations expressing these properties hold on the dyadic simple functions. By continuity, the restriction of Φ to the continuous functions that depend on only one coordinate agrees with the Borel functional calculus for that T_n. In particular, $\Phi(x_n) = T_n$. This completes the proof of the first statement.

For the second statement, recall from Theorem 3.60 that $\ker \Phi$ is the set of continuous functions vanishing on some closed subset $K \subseteq [-1, 1]^N$ and that $C([-1, 1]^N)/\ker \Phi \cong C(K)$ by the natural map. So Φ induces a nonexpansive, injective, unital, self-adjoint homomorphism from $C(K)$ into $\mathcal{B}(H)$.

We claim that for any k, any point $x \in K$, and any open neighborhood U of x, there is a cube A_i of side length 2^{-k} that intersects U and satisfies $\Phi(1_{A_i}) \neq 0$. If this claim were to fail then Φ would take any continuous function that vanishes off of U to 0 (since any such function is uniformly approximated by dyadic simple functions supported on cubes which intersect U), contradicting injectivity on $C(K)$. Thus the norm computation from Proposition 5.31 shows that $\|\Phi(f)\| \geq |f(x)|$ for all $f \in C([-1, 1]^N)$. As this holds for all $x \in K$, we get $\|\Phi(f)\| \geq \|f|_K\|_\infty$. So the induced map from $C(K)$ into $\mathcal{B}(H)$ is isometric. □

A C*-algebra $\mathcal{A} \subseteq \mathcal{B}(H)$ is *nondegenerate* if the set $\{Tv : T \in \mathcal{A}, v \in H\}$ is dense in H. Every C*-algebra in $\mathcal{B}(H)$ is the direct sum of a nondegenerate C*-algebra in $\mathcal{B}(E)$ with the C*-algebra $\{0\}$ in $\mathcal{B}(E^\perp)$, for some closed subspace $E \subseteq H$ (exercise).

Theorem 5.45. *Let H be a separable complex Hilbert space and let $\mathcal{A} \subseteq \mathcal{B}(H)$ be a nondegenerate separable abelian C^*-algebra. Then there is a metrizable locally compact Hausdorff space X, a finite, strictly positive Borel measure μ on X, a measurable Hilbert bundle $\mathcal{H}$ over X, and a Hilbert isomorphism $U : H \cong L^2(X, \mu; \mathcal{H})$ such that the map $f \mapsto U^{-1}M_f U$ is an isometric self-adjoint homomorphism from $C_0(X)$ onto $\mathcal{A}$.*

Proof. Let (T_n) be a sequence of operators whose span is dense in $\mathcal{A}$. By the comment just after Definition 5.43, we can assume the T_n are self-adjoint. We may also assume that $\|T_n\| < 1$ for all n.

By the lemma, for each N we have a nonexpansive, unital, self-adjoint homomorphism $\Phi_N : C([-1, 1])^N \to \mathcal{B}(H)$ taking the coordinate function x_n to the operator T_n for $1 \leq n \leq N$ and with kernel the continuous functions vanishing on some closed subset $K_N \subseteq [-1, 1]^N$. Since these maps are all homomorphisms, $\Phi_{N'}$ must agree with Φ_N on all polynomials in $x_1, \ldots, x_N$ for any $N' > N$. So we can consistently define a map Φ from all polynomials in the variables x_n into $\mathcal{B}(H)$ which takes x_n to T_n, and by continuity it extends to a nonexpansive, unital, self-adjoint homomorphism from $C([-1, 1]^{\mathbf{N}})$ to $\mathcal{B}(H)$. By Theorem 3.60, $\ker \Phi$ is the set of continuous functions vanishing on some closed subset $K \subseteq [-1, 1]^{\mathbf{N}}$ and we have $C([-1, 1]^{\mathbf{N}})/\ker \Phi \cong C(K)$ by the natural map. Embedding $C([-1, 1]^N)$ into $C([-1, 1]^{\mathbf{N}})$ as the continuous functions which only depend on the first N coordinates, we also have that $\Phi_N = \Phi|_{C([-1,1]^N)}$, and therefore K_N is the projection of K onto $[-1, 1]^N$. (For $f \in C([-1, 1]^N)$, we have $\Phi(f) = 0$ if and only if f vanishes on the projection of K, while $\Phi_N(f) = 0$ if and only if f vanishes on K_N. Since both K_N and the projection of K are closed, this implies that they must be equal.) It follows that the sup norm of any $f \in C([-1, 1]^N)$ restricted to K_N equals its sup norm, regarded as a function on $[-1, 1]^{\mathbf{N}}$, restricted to K. Thus $\Phi : C(K) \to \mathcal{B}(H)$ is isometric on those functions which depend only on the first N coordinates, for all N; by Stone-Weierstrass these functions are dense in $C(K)$, so Φ is isometric on all of $C(K)$.

Applying Lemmas 5.39 and 5.41, we get a finite, strictly positive Borel measure μ on K, a measurable Hilbert bundle $\mathcal{H}$ over K, and a Hilbert isomorphism $U : H \cong L^2(K, \mu; \mathcal{H})$ such that $\Phi(f) = U^{-1}M_f U$ for all $f \in C(K)$.

Since the operators T_n become multiplication by coordinate functions in the $L^2(K, \mu; \mathcal{H})$ picture, nondegeneracy implies that if the point $\mathbf{0} = (0, 0, \ldots)$ belongs to K then $\mu(\{\mathbf{0}\}) = 0$. So we can replace K with $X =$

$K \setminus \{0\}$ and still have $\Phi(f) = U^{-1}M_f U$ for all $f \in C_0(X)$. Then $\Phi(C_0(X))$ clearly contains $\mathcal{A}$, since Φ is isometric and it contains each T_n. Conversely, $\Phi(p) \in \mathcal{A}$ for any polynomial p in the coordinate functions with no constant term, and as these are dense in $C_0(X)$ we conclude that Φ takes $C_0(X)$ isometrically onto $\mathcal{A}$. $\square$

The next and final result naturally closes up this circle of ideas. The Riesz-Markov theorem already showed us that there is a surprisingly fundamental connection between topology and measure theory. We can now see that topology, measure theory, and Hilbert space are all deeply interlinked.

Theorem 5.46. *Let H be a separable complex Hilbert space and let $\mathcal{M} \subseteq \mathcal{B}(H)$ be a nondegenerate abelian von Neumann algebra. Then there is a metrizable locally compact Hausdorff space X, a finite, strictly positive Borel measure μ on X, a measurable Hilbert bundle $\mathcal{H}$ over X, and a Hilbert isomorphism $U : H \cong L^2(X, \mu; \mathcal{H})$ such that the map $f \mapsto U^{-1}M_f U$ is an isometric, weak* homeomorphic, self-adjoint homomorphism from $L^\infty(X, \mu)$ onto $\mathcal{M}$.*

Proof. Find a sequence of operators (T_n) in $[\mathcal{M}]_1$ which is weak* dense in $[\mathcal{M}]_1$. Then the norm closure of the polynomials in the operators T_n and their adjoints is a separable abelian C*-algebra $\mathcal{A}$ which is weak* dense in $\mathcal{M}$. Since $\mathcal{M}$ is nondegenerate, $\mathcal{A}$ must also be nondegenerate. (If the set $\{Tv : T \in \mathcal{A}, v \in H\}$ were not dense in H, then there would be a nonzero vector $w \in H$ such that $\langle Tv, w \rangle = 0$ for all $T \in \mathcal{A}$ and $v \in H$. Then by weak* density this would also be true for all $T \in \mathcal{M}$, contradicting nondegeneracy of $\mathcal{M}$.)

Thus we can apply Theorem 5.45 to get a metrizable locally compact Hausdorff space X, a finite, strictly positive Borel measure μ on X, a measurable Hilbert bundle $\mathcal{H}$ over X, and a Hilbert isomorphism $U : H \cong L^2(X, \mu; \mathcal{H})$ such that the map $f \mapsto U^{-1}M_f U$ is an isometric self-adjoint homomorphism from $C_0(X)$ onto $\mathcal{A}$. The map $f \mapsto U^{-1}M_f U$ remains isometric on $L^\infty(X, \mu)$ since $\|M_f\| = \|f\|_\infty$, as we noted at the beginning of Section 5.4. It is clearly a self-adjoint homomorphism. To verify weak* continuity, let (f_n) be a bounded sequence in $L^\infty(X, \mu)$ that converges weak* to $f \in L^\infty(X, \mu)$. Then for any $\eta, \xi \in L^2(X, \mu; \mathcal{H})$ we have

$$\langle M_{f_n}\eta, \xi \rangle = \int f_n(x)\langle \eta(x), \xi(x) \rangle;$$

but $|\langle \eta(x), \xi(x) \rangle| \leq \|\eta(x)\|\|\xi(x)\|$, and the product of two functions in $L^2(X, \mu)$ belongs to $L^1(X, \mu)$ by Hölder's inequality, so the function $x \mapsto$

$\langle \eta(x), \xi(x) \rangle$ belongs to $L^1(X, \mu)$ and we therefore have

$$\langle M_{f_n} \eta, \xi \rangle \rightarrow \langle M_f \eta, \xi \rangle.$$

This shows that $f \mapsto U^{-1} M_f U$ is weak* continuous. Then the unit ball of the image of $L^\infty(X, \mu)$, which is also the image of the unit ball, is weak* compact, and hence weak* closed, so the image of $L^\infty(X, \mu)$ is weak* closed by Krein-Smulian. Thus the image of $L^\infty(X, \mu)$ contains $\mathcal{M}$. Conversely, the image of $L^\infty(X, \mu)$ is contained in the weak* closure of $\mathcal{A}$ by Proposition 4.46 and weak* continuity. So the image of $L^\infty(X, \mu)$ must equal $\mathcal{M}$.

The inverse map from $\mathcal{M}$ to $L^\infty(X, \mu)$ is weak* continuous by the same argument as in the proof of Lemma 5.39. $\qquad\qquad \square$

5.7 Exercises

5.1. Assuming real scalars, show that l_2^1 is isometrically isomorphic to its dual. Is this also true when the scalars are complex?

5.2. Let v and v' be vectors in a Hilbert space H and suppose $\langle v, w \rangle = \langle v', w \rangle$ for all $w \in H$. Show that $v = v'$. What if H is equipped with an inner product but we do not assume completeness?

5.3. Let v and v' be vectors in a Hilbert space and suppose $\|v\| = \|v'\|$. Prove that there is a Hilbert isomorphism from H to itself that takes v to v'. Can we interchange v and v'? What if the norm on H comes from an inner product but we do not assume completeness?

5.4. Let H be a separable Hilbert space and let $E \subseteq H$ be a dense, countable-dimensional subspace. Prove that there exists an orthonormal basis of H whose span equals E.

5.5. Let (X, μ) be a σ-finite measure space. Show that $L^2(X, \mu)$ is separable iff (X, μ) is measurably separable.

5.6. Let (X, μ) be a measurably separable σ-finite measure space and let $\mathcal{H}$ be a measurable Hilbert bundle over X. Prove that $L^2(X, \mu; \mathcal{H})$ is separable.

5.7. Let E and F be closed subspaces of a Hilbert space. Prove that $P_E P_F = P_E$ iff $E \subseteq F$.

5.8. Characterize which closed subspaces E and F of a Hilbert space satisfy $P_E P_F = P_F P_E$.

5.9. Let (X, μ) be a σ-finite measure space and let H be a separable Hilbert space. Suppose $\{U_x : x \in X\}$ is a family of Hilbert isomorphisms of H with itself such that for all $v, w \in H$ the function $x \mapsto \langle U_x v, w \rangle$ is measurable. Prove that the map taking $\eta(x)$ to $U_x \eta(x)$ is a Hilbert isomorphism of $L^2(X, \mu; H)$ with itself that restricts to a Hilbert isomorphism of $L^2(A, \mu|_A; H)$ with itself, for all measurable $A \subseteq X$.

5.10. Let (X, μ) be a σ-finite measure space and let H be a separable Hilbert space. Suppose U is a Hilbert isomorphism of $L^2(X, \mu; H)$ with itself that restricts to a Hilbert isomorphism of $L^2(A, \mu|_A; H)$ with itself, for all measurable $A \subseteq X$. Prove that U has the form described in Exercise 5.9. (Look at the images of the functions $1_X \cdot e_n$ under U, where (e_n) is an orthonormal basis of H.)

5.11. For any bounded Hilbert space operator T, prove that $\ker T = T^*(H)^\perp$.

5.12. Let E and F be closed subspaces of a Hilbert space H. An operator $U \in \mathcal{B}(H)$ is a *partial isometry* from E to F if U restricts to a Hilbert isomorphism between E and F and $U|_{E^\perp} = 0$. Prove that U is a partial isometry iff U^*U is a projection iff UU^* is a projection.

5.13. Let (X, μ) be a σ-finite measure space and let H be a separable Hilbert space. A *weakly measurable field of operators* on X is a map $\mathcal{T} : X \to \mathcal{B}(H)$ such that $x \mapsto \langle \mathcal{T}(x)v, w \rangle$ is measurable for all $v, w \in H$. Prove that any bounded, weakly measurable field of operators $\mathcal{T}$ defines an operator $T \in \mathcal{B}(L^2(X, \mu; H))$ by $(T\eta)(x) = \mathcal{T}(x)\eta(x)$. Prove that $L^2(A, \mu|_A; H)$ is invariant for T, for every measurable $A \subseteq X$.

5.14. In the notation of Exercise 5.13, let T be a bounded operator on $L^2(X, \mu; H)$ for which $L^2(A, \mu|_A; H)$ is invariant, for every measurable $A \subseteq X$. Prove that T arises from a bounded, weakly measurable field of operators on X in the manner described in Exercise 5.13. (Look at the images of the functions $1_X \cdot e_n$ under T, where (e_n) is an orthonormal basis of H.)

5.15. Prove that any closed convex subset of a Hilbert space contains a unique element of minimal norm. (Fix a sequence of points in the set whose norms decrease to the minimal value and use the parallelogram law to show that the sequence must be Cauchy.)

5.16. (Polarization identity) Let H be a complex Hilbert space, let $S \in \mathcal{B}(H)$, and let $v, w \in H$. Prove that

$$\langle Sv, w \rangle = \frac{1}{4}\big(\langle S(v + w), v + w \rangle + i\langle S(v + iw), v + iw \rangle$$
$$- \langle S(v - w), v - w \rangle - i\langle S(v - iw), v - iw \rangle\big).$$

5.17. Suppose $U : H \to K$ is a surjective linear isometry between two complex Hilbert spaces, and prove that U is a Hilbert isomorphism. (Use polarization.) What about the real case?

5.18. A self-adjoint operator $T \in \mathcal{B}(H)$ is *positive* if $\langle Tv, v \rangle \geq 0$ for all $v \in H$. Prove that if $\mathbf{F} = \mathbf{C}$ then the condition $\langle Tv, v \rangle \geq 0$ for all v implies that T is self-adjoint. (Use polarization.) What about the real case?

5.19. Let (X, μ) be a σ-finite measure space, let $\mathcal{H}$ be a measurable Hilbert bundle over X, and let $f \in L^{\infty}(X, \mu)$. Prove that M_f is positive (see Exercise 5.18) iff $f(x) \geq 0$ almost everywhere.

5.20. Assume $\mathbf{F} = \mathbf{C}$ and identify $\mathcal{B}(\mathbf{C}^2)$ with the set of 2×2 complex matrices. Prove that every self-adjoint operator can be written as a real linear combination of the identity matrix and the *Pauli matrices*

$$\begin{bmatrix} 1 & 0 \\ 0 & -1 \end{bmatrix}, \quad \begin{bmatrix} 0 & 1 \\ 1 & 0 \end{bmatrix}, \quad \begin{bmatrix} 0 & i \\ -i & 0 \end{bmatrix}.$$

Prove that the set of real linear combinations of the Pauli matrices, with operator norm, is isometrically isomorphic to a three-dimensional real Hilbert space. (Use $\|T^*T\| = \|T\|^2$ to compute norms.)

5.21. Let T be a positive operator on a separable Hilbert space H (see Exercise 5.18). Prove that $\mathrm{sp}(T) \subset [0, \infty]$.

5.22. For finite-dimensional H, interpret Theorem 5.42 in terms of eigenvalues and eigenvectors.

5.23. Let H be a separable Hilbert space. Prove that T^*T is positive (see Exercise 5.18) for any $T \in \mathcal{B}(H)$. Prove that every positive operator equals T^*T for some $T \in \mathcal{B}(H)$. Can we ensure that this T is self-adjoint?

5.24. Let H be a separable Hilbert space and let $T \in \mathcal{B}(H)$ be self-adjoint. Suppose $\mathrm{sp}(T) \subseteq [-\epsilon, \epsilon] \cup [1 - \epsilon, 1 + \epsilon]$. Prove that there is a projection P such that $\|T - P\| \leq \epsilon$.

5.25. Let H be a separable Hilbert space, let $T \in \mathcal{B}(H)$ be self-adjoint, and suppose T is invertible in $\mathcal{B}(H)$. Find a map $f : [0, 1] \to \mathcal{B}(H)$, continuous for the norm topology on $\mathcal{B}(H)$, such that $f(0) = I$, $f(1) = T$, and $f(t)$ is invertible for all t. Can we also ensure that $f(t)$ is self-adjoint for all t?

5.26. Let H be a separable compled Hilbert space and let $\mathcal{A} \subseteq \mathcal{B}(H)$ be a C*-algebra. Prove that there is a closed subspace $E \subseteq H$ such that $T(E) \subseteq E$ for all $T \in \mathcal{A}$, $T|_{E^\perp} = 0$ for all $T \in \mathcal{A}$, and $\{Tv : T \in \mathcal{A}$ and $v \in E\}$ is dense in E.

5.27. Let H be a separable complex Hilbert space and let $\mathcal{M} \subseteq \mathcal{B}(H)$ be an abelian von Neumann algebra. Prove that there is a closed subspace $E \subseteq H$ such that $P_E \in \mathcal{M}$ and $T|_{E^\perp} = 0$ for all $T \in \mathcal{M}$.

5.28. Let H be a separable complex Hilbert space and let $T \in \mathcal{B}(H)$. Suppose $TT^* = T^*T$ (T is *normal*). Prove that there is a compact subset X of $\mathbf{C}$, a finite, strictly positive Borel measure μ on X, a measurable

Hilbert bundle $\mathcal{H}$ over X, and a Hilbert isomorphism $U : H \cong L^2(X, \mu; \mathcal{H})$ such that $T = U^{-1}M_zU$. (Decompose T into real and imaginary parts.)

5.29. Let V be a unitary operator on a separable complex Hilbert space H and let $\mathbf{T} = \{z \in \mathbf{C} : |z| = 1\}$ be the unit circle in the complex plane. Prove that there is a finite positive Borel measure on $\mathbf{T}$, a measurable Hilbert bundle $\mathcal{H}$ over $\mathbf{T}$, and a Hilbert isomorphism $U : H \cong L^2(\mathbf{T}, \mu; \mathcal{H})$ such that $V = U^{-1}M_zU$.

5.30. Let H be a separable complex Hilbert space, let $\mathcal{A} \subseteq \mathcal{B}(H)$ be a C*-algebra, and let $T \in \mathcal{A}$ be self-adjoint. Suppose T is invertible in $\mathcal{B}(H)$. Prove that T^{-1} lies in $\mathcal{A}$.

5.31. Let H be a separable complex Hilbert space, let $\mathcal{M} \subseteq \mathcal{B}(H)$ be a von Neumann algebra, let $T \in \mathcal{M}$ be self-adjoint, and let $E = \overline{T(H)}$. Prove that P_E (the *range projection* of T) lies in $\mathcal{M}$.

5.32. Let (X, μ) be a measurably separable σ-finite measure space and let $\mathcal{H}$ be a measurable Hilbert bundle over X. Prove that $\{M_f : f \in L^\infty(X, \mu)\}$ is an abelian von Neumann algebra in $\mathcal{B}(L^2(X, \mu; \mathcal{H}))$.

5.33. A *masa* in $\mathcal{B}(H)$ is an abelian von Neumann algebra which is not properly contained in any other abelian von Neumann algebra. Prove that the von Neumann algebra constructed in Exercise 5.32 is a masa iff the fiber of $\mathcal{H}$ over almost every point is one-dimensional.

5.34. Let H be a separable Hilbert space and let $\mathcal{M} \subseteq \mathcal{B}(H)$ be a von Neumann algebra. Assume $I \in \mathcal{M}$. The *center* of $\mathcal{M}$ is the set $Z(\mathcal{M}) = \{T \in \mathcal{M} : ST = TS \text{ for all } S \in \mathcal{M}\}$. Prove that $Z(\mathcal{M})$ is an abelian von Neumann algebra. Apply Theorem 5.46 to $Z(\mathcal{M})$ and show that $L^2(A, \mu|_A; \mathcal{H}|_A)$ is invariant for UTU^{-1}, for all measurable $A \subseteq X$ and all $T \in \mathcal{M}$.

Notation Index

197

Subject Index